OUR QUANT
AND REINCARNATION
VOLUME II

(SOMETHING SURVIVES)

MILTON E. BRENER

Printed in the United States of America

ISBN: Softcover   978-1-63871-242-8
         Harback   978-1-63871-362-3
         eBook   978-1-63871-243-5

Republished by: PageTurner Press and Media LLC
Publication Date: 07/27/2021

To order copies of this book, contact:

PageTurner Press and Media
Phone: 1-888-447-9651
info@pageturner.us
www.pageturner.us

To Eileen

Chapter 1:

INTRODUCTION

Thisis a supplement to my book published in 2015 entitled Our Quantum World and Reincarnation. It is also a compendium of blogs published in my web site at rebirthandquantum.com and in blogger.com, an appendage of Google. Before talking about the reasons for this essay, a much abbreviated recap of the subject itself might be useful.

The subject is reincarnation, though probably not the reincarnation envisioned by most believers. My materiel for part of it is primarily the result of the lifelong work of the physician and scholar, Ian Stevenson and others who have followed in his footsteps. He died in 2007, having reported his investigations of over 2500 cases worldwide of children born with memories of a previous life.

Most of them spoke about details of the previous life beginning with their earliest speech, sometimes about 2 or 3 years of age. It often included, increasingly in the following few years, distant towns or cities, with descriptions of houses, including interiors, layout of buildings, and names and identity of persons. Stevenson and his co-workers traveled extensively to view physical evidence and to question and record the statements of witnesses.

The researchers always sought corroboration from others as to the correctness of the recollections of the subjects, mostly children. Those reports by subjects accepted as genuine by the investigators were thus backed by a preponderance of proof, including statements of parents, other relatives, and family friends. There was additional evidence or apparent circumstance that there was no way for the child to have known of such facts through normal channels.

To Stevenson, the person reporting the memories is the 'subject'. The person, when known, whose former life was remembered, is the 'prior personality.'

In most cases, though not all, these memories of the children seem to diminish by ages 7 or 8 with their immersion in current surroundings and current lives. Ultimately they often disappear. In the following pages 13 of these cases are summarized. That includes five summarizations in my previous book, namely the three cases stemming from WWII, Chapter 14 of this essay, and the cases of Marta Lorenz and Corliss

Chotkin, Jr., Chapter 12. Not all of these 13 were investigated by Stevenson. Even many who disagree with Stevenson's conclusions praise his method and the scientific nature of his investigation.

Our other subject is 'entanglement,' one aspect of quantum physics. It is of primary interest as a possible explanation of this strange phenomenon of reincarnation, or at least the limited form of it shown by investigation. I will refer to it as reincarnation as that is the term used by Dr. Stevenson and other investigators. I tend to see it however as a third layer of inheritance, the first two being genetics and epigenetics, a term to be explained later, in Chapter 6.

Entangled particles have characteristics complementary to each other. Spin, for instance, is a characteristic of electrons. If one of two entangled electrons spins right, we would expect the other of the pair to spin left.

Once the entanglement between two or more atoms or particles is formed, it will, for our purposes, last forever. Also, once formed, the relationship continues no matter how far any subsequent separation between them may be, even millions or billions of light years. Each light year consists of almost six trillion miles, the distance light can travel in a year.

The change in one particle is accompanied by a change in the other so as to keep the complementary relationship intact. The change is instantaneous, each changing simultaneously with the other.

If you find, as does almost everyone, that quantum physics, including the aspect of it known as entanglement, is baffling and hard to understand, you should hear from Richard Feynman. He was an American theoretical physicist who died in 1988 at the age of 70. In 1965 he was awarded a Nobel Prize for his work with subatomic particles, and in a 1999 poll of scientists by the British journal *Physics World* he was ranked as one of the ten greatest physicists of all time. More interesting perhaps is a characterization of him as a "Nobel Laureate, professor, musician, raconteur and an all-around curious character."

The last two items in that description are well illustrated in a public lecture he delivered on quantum physics: "You think I'm going to explain it to you so you can understand it?" he asked the audience. "No you're not going to be able to understand it . . . You see my physics students don't understand it either. This is because I don't understand it. Nobody does."

It is almost as much a mystery to scientists today as it was during his lifetime.

There are a number of pithy sayings about quantum bandied about by so many scientists in so many different ways that it is often difficult to know who was first to say any of them. One of the more frequent ones is to the effect that anyone who claims to understand quantum physics doesn't know what he is talking about.

Yet many accomplished and well respected scientists say in speeches and in their writings, that laymen should not be put off or frightened away by the term, 'quantum.' Among them:

Bruce Rosenblum and Fred Kuttner have written: *It does not require a technical background to move to the frontier where physics join issues that seem beyond physics."* It is a realm, they claim, where *"physicists cannot claim complete competence,"* and that *"once there you* [the laymen] *can take sides in the debate;* **Michael Talbot** wrote that his experience had taught him not only that those who do not know any mathematics are able to understand the kinds of ideas he advances, but that *You do not even need a background in science;* **Brian Clegg** wrote: *I quite happily teach the basics of physics to ten-year-olds. Not the maths, but you don't need mathematics to appreciate what's going on. You just need the ability to suspend your disbelief. Because quantum particles refuse to behave the way you'd expect;* and **Nicolas Gisin** wrote that *Mathematics is not needed to tell the great story of physics. For what is interesting in physics is not the mathematics but the concepts. . . . At the end of the day, you will find that one can understand quite a lot of quantum physics without the need for mathematics.*

All of these opinions, and many others, considered, I see no reason why I should not advance the ideas here to be detailed, or why you should not understand them, at least to the degree that anything about them can yet be understood.

The ultimate thesis of my book and this essay is the clear possibility that the atoms, the eternal atoms that form the brain organs dealing with memories, are entangled, and can and do communicate with each other even after the death of the 'prior personality' and in some cases can and do retain the memories of that deceased person. It further involves a possibility that they may in some instances form part of the brain of a fetus or newborn. The result may be the phenomenon that Stevenson and others have investigated. In short, we are all made of atoms, and the atoms do survive us. Those are facts. The laws of quantum physics are also fact. The balance is my thesis, offered as a possibility.

I have never claimed, and do not claim now, that I have proved the proposition that those facts account for the other fact, namely the memories of some youngsters just described. I can claim at most that the accepted facts of quantum physics *may*, in the present state of knowledge, account for it. Though I cannot prove that it does, as far as I know it could not at present be proved that it cannot. One day new discoveries may prove the thesis is impossible; but may likewise prove that it is in fact possible. Or the entire idea may simply die an isolated, and unnoticed death.

So, why do I write this essay? Mainly because since publication of my book, there has been, and continues to be, a tremendous amount of research and progress in the field of quantum physics and related areas. Some of it I found to be very relevant in support of my ideas. Some of it was available to the public only after I had submitted the final draft of that book for publication. Other material was very much available, such as the writings of Lawrence Krauss and Primo Levi, summarized in Chapter 9, but appeared to me unnecessary as they dealt with the life of atoms of oxygen and carbon respectively. These are two of the four elements absolutely vital to human and most other animal life and both seem to be eternal.

Krauss dealt with, and emphasized the fact that we breathe the air full of atoms of oxygen previously breathed by others throughout human history and its forebears. Levi did the same with carbon and the plants, humans and other animals, rich in atoms of that element. I thought I had covered sufficiently a clear possibility that fetuses or young children could be the recipients of entangled particles from the organs of memory of prior personalities. I had also spelled out, I believe, the fact that many atoms, from our perspective, last 'forever,' or, at least, billions of times longer even than the life of our solar system. I thought that that aspect of the subject would need little further explanation.

Until, that is, I read the words of one reviewer who faulted me for not explaining how the result in question could occur. I realized there may be others who felt the same lack. Hence I have added in this essay, a summary of the writings of these two scholars though neither of them, Krauss or Levi, were dealing with reincarnation, or quantum physics.

On another subject, anyone taking stock of the statistics in this area would have noticed the relatively few cases of reports of recollection of events or people, or anything enlightening, occurring during the period between the death of the prior personality and the birth of the subject person.

I later delved again into the subject of current thinking about the location and processes involved in the forming, or coding, of memories; storage of them, and retrieval with this problem in mind. The entire subject is very complex, and much is still unknown, but it is clear that the encoding and retrieval are far more complex than the storage. It seems doubtful that the 'period in between' could be productive of new thoughts that could be remembered. What can be remembered by the subject are thoughts and memories stored during the life of the prior personality. It is a subject highly relevant to our thesis here and chapter 11 is devoted, in part, to it.

I have also added a chapter on 'regression therapy.' The purpose of this procedure is healing of the patient, often beginning with hypnosis when all else has failed, to determine whether there exists unconscious memories of another life, and their connection, if any, with current symptoms. Because of the difference in purpose from the cases in other chapters, the validity or accuracy of the memories in this chapter is not of significant importance. Nonetheless, in a few cases, including the one summarized here in Chapter 8, there is proof of both validity and accuracy of accounts under hypnosis of a prior life, which did certainly cast light upon a phobia in this life, and resulted in substantial healing.

I have included also summaries of cases, some not used in my book, and pairings of them which I believe highlight similarities and differences between various cases which come from different lands and cultures in all parts of the world.

One line of thought now given serious consideration by physicists is the atom's possible function in holding together all of space. It is entanglement say these physicists that stitches the universe together.

Others disagree with most of it. You and I are entitled to do the same. But agree or disagree there is much to chew on, and perhaps to find extraordinarily interesting.

Those are the major additions. There are other minor ones. Other subjects such as philosophy and ancient religions that I touched on in my book have been omitted here. In this essay, we will stick to science, and leave the remainder to rest in peace in the mentioned book of mine and the many books of many others. I hope you will find it, or at least some of it, as interesting as I do.

Chapter 2:

THE CASE OF EINAR JONSSON

Before getting into any discussion of quantum physics, and entanglement we should see at least one sample of the work of Ian Stevenson. Of the 2500 or so that Stevenson reported, I have read a little over 100. The ones I will summarize, like the 12 in my recent book, are not selected as being either the most typical or the best. I believe however that they typify the universal character of this phenomenon, and the diversity of experiences inherent in them. Not all of them are by Stevenson.

My summaries, it should be understood, are much shortened versions of the thoroughly documented originals by Dr. Stevenson and others.

The subject of this case, **Einar Jonsson**, was born on July 26, 1969 to Jon Nielsson (Jon) and Helga Haraldsdottir (Helga) in Reykjavik, Iceland. In Iceland the wife does not take the name of her husband. At about 18 months Einar first began to speak and at about 2 years of age he began to speak of a man on a tractor who died and of a different mother who, he said, also died. He gradually spoke also about a farm with cows, sheep and a boat. The family wondered if his statements referred to a youth named Harald Olafsson (Harald) who had died in a tractor accident on July 18, 1969, eight days before Einar's birth. In Iceland, as in most of the world's cultures, Western Europe and the lower 48 states of the United States excepted, reincarnation is accepted as a frequently or sporadically appearing fact. There is no stigma attached to such belief or to open mention and discussion of it.

An associate of Dr. Stevenson, Dr. Erlendur Haraldsson, interviewed Einar's mother, Helga, in November 1973. Einar himself however had stopped talking about a prior life at about age four and would not respond to any of the investigator's questions.

Stevenson interviewed both parents in 1980, and in 1985 interviewed Einar's paternal grandmother, Marta Sigurdsdottir (Marta). Marta was the mother of Jon, Einar's father by her first husband, Neils Larusson (Neils). By a second marriage to Olaf Petursson (Olaf), she was also the mother of Harald, the suspected prior personality. Harald was born in May 1955, and was the victim of the tractor accident.

Hence Jon, Einar's father, and Harald were half-brothers, namely same mother, different fathers. Einar, the 'subject' claiming a prior life, was the nephew of Harald, the prior personality. On July 18, 1969 Harald was returning from a neighbor's farm on a family tractor when it went off the road and he was killed instantly. This, as previously stated, was eight days before the birth of Einar.

Now, at the 1985 meeting, almost 16 years old, and willing to discuss his memories, Einar, and his mother, Helga, were present and added certain information. Several of Einar's statements in his early months and years, included the house where Marta's son Harald had lived. These needed verification, which required Stevenson and his assistant travelling to the town of Laufas, about 4 ½ hours by car, which they did in October 1999. In addition to those already mentioned, they interviewed Olaf, the father of Harald, who could furnish little information except as mentioned below.

Marta and Neils, with their son Jon, lived in Reykjavik until Jon was about 5 years old. She then moved to Laufas where she married Olaf, who owned a farm. Four more sons were born to Marta in that area of which Harald was the second. Jon remained at Laufas until about 16 or 17 years of age.

Thus Jon, the father of Einar, had spent about 12 years at Laufas and knew the area well. He was also well acquainted with his half-brother Harald who was about 8 years younger than he. Harald had been about 9 years old when Jon left Laufas. Jon was a principal verifier of Einar's statements. His wife Helga, Einar's mother, however never knew Harald and she had never been to Laufas until she took Einar there to spend the summer at the age of five.

In his 1999 journey to Laufas Stevenson was armed with 16 statements known to have been made by Einar between the ages of two and four. Helga was certain that all of the statements made by Einar were between those ages, a year before he spent a summer at Laufas. Stevenson found 13 of the statements concerning the farm, its surroundings, and some family relationships to be correct; 3 were incorrect, and it appears that, claiming the death of his 'other' mother, can be easily explained. Marta was still alive, but Einar never met her, and when he made his 'incorrect' statement, from his perspective she was gone.

Einar had correctly said that the man who died, Harald, had a big brother. He also said there was another man who limped. This was, in fact, Harold's maternal grandfather. He had stayed in the home for six months during the last year of his life when Harold was 3 or 4 years old, and he did in fact limp. The farmhouse was large. There was a mountain behind the house with an unusual shape, which Stevenson described as conical, rather than flat topped like most mountains in the area. There was a small boat on the farm but it had become 'broken.' Stevenson was shown a picture of the boat in its destroyed condition which resulted from a storm.

There was, according to Einar, a cow house and sheep house that burned, but no one remembered any such fire. Einar said that Harald had skis, but, according to Helga, no one in the family had skis. Einar also claimed, as already explained, he had another mother who was dead. Verification of the correct statements came from Jon, Marta, and/or Helga and their personal observations. Olaf was also able to confirm, with Jon, the 'broken' boat.

Helga was asked by Stevenson if, when the five year old Einar spent the summer at Laufas, he showed any signs of familiarity with the farm and its area, which would have presumably come from his 'prior life' as Harald. She replied that he did not, but that she herself had. This occurred, she thought, because of the accuracy of Einar's description of the large farmhouse and the oddly shaped mountain behind it.

Unlike most children who have memories of another life, he narrated more often in the third person rather than the first person, more like an observer watching events than living them, and about a deceased person rather than being one. But he had rejected his mother, Helga, for a time and would not let her touch him.

Stevenson's final comment on the case: "Tractor accidents are common on farms; fatal accidents much less so. Many statements Einar made about the farm would be correct for many farms, perhaps for nearly all. Yet only farms near water would have a boat, and not many of these would have a man in the house who limped. Also not many farms would have a mountain of unusual shape just behind them. . . . We cannot attach an estimate of probability to this group of details, but I think most readers will agree with me that we are unlikely to find them all together in farms other than the ones where Harald Olafsson lived."

He specifically agreed that the statements of Einar referred to a particular youth, Harald, the stepbrother of the subject's father. The expressed conclusion places this case squarely within Stevenson's defined group of cases he considers 'solved.' Nonetheless, he calls it a "guarded affirmative."

Though later continuing with more samples of this mysterious type of occurrence, we should next turn to some very basic material about those parts of quantum physics we may need to better understand entanglement and its possible role in cases such as this.

Chapter 3 :

THE MIGHTY ATOMS

Having offered a small, and hopefully tantalizing, peek at the work of Ian Stevenson and his colleagues in the prior chapter, I would like to offer the same with regard to the other mystery incorporated in my book *Our Quantum World and Reincarnation*. That, of course is quantum physics, often called for some reason 'quantum mechanics.' Much more qualified persons than I have preferred 'quantum physics,' so I feel justified in sticking with it. By whatever name, it is full of mind blowing numbers, large and small.

Atoms, including its sub-atomic particles, are essentially what the world, including humans, are made of. Just how tiny are the atoms? Its size can be described as the number of them in a thimble, which is 10 followed by @24 zeroes. Perhaps more intelligible to us is the description by one scientist who has made the study. He claims that there are as many atoms in a drop of water, as there are drops of water in all the lakes and rivers [though not the oceans], on our planet.

There are other even tinier particles. Apart from those forming components of the atom, there are photons, not to be confused with protons, a component of the nucleus. The photons are the carriers of light and all its relatives such as electromagnetism. According to most physicists, photons have no mass. Light and electromagnetism permeate the universe and everything in it, including our bodies, and most particularly from our viewpoint, our brains. Electro-chemicals provide the connections, called synapses, among our brains' 82 billion neurons, which are the cells of the brain. Electromagnetism is as vital to the existence of the universe, including ourselves, as are the atoms and their parts.

Then there are mysterious particles known as neutrinos, one of the most common fundamental particles in the universe. They are abundantly produced in various nuclear processes in space involving the decay of radioactive elements. Trillions of neutrinos pass through each of our bodies every minute. Yet, they are very difficult to study because they rarely interact with their surroundings and thus often evade detection.

They are the smallest particle known and are estimated to have a mass of two millionths the mass of the next low mass particle, the electron, which, in turn is only 1/1837 as heavy as the mass of the proton. It has been responsibly claimed that statistically each neutrino could pass through solid lead, many light years thick, each light year being almost 6 trillion miles, without encountering any atom or other particle. As the name implies they have no electric charge. They must have some effect on something somewhere, but so far no one seems to know what, why, or where.

But the actual structure of the universe and everything in it are the atoms and their parts. One reason the neutrinos so rarely hit atoms or other particles is that the atoms, are mostly empty space. We temporarily pass over the proposition that space may not be really empty. Most high school graduates know that atoms consist of a nucleus and surrounding rings of electrons. Since the advent of quantum physics, recognizing that the exact position of an electron cannot be determined, the rings are better known as electron clouds, or as possibility waves, about which we will later hear more.

What the high school student may not know is that that if the nucleus, which is about one trillionth of a centimeter, were as large as a golf ball, the first electron cloud would be about .62 or 1.5 miles away, depending on which source one prefers. The electron itself, which is two million times more massive than the neutrino, is only about one sixteen hundredth as massive as the nucleus of the atom.

The word 'atom' comes from a Greek word meaning indivisible. It was said by the Ancient Greeks to be the very smallest bit of matter, and constituted its basic structure. It was a remarkable insight, but erroneous as proved by modern science. Not only have atoms been found to have electrons and nucleus, but the nucleus consists of parts called protons and neutrons, which, in turn, are made up of 'quarks.'

The major difference between atoms is their 'atomic number,' a reference to the number of protons and of neutrons, which are almost always the same number as the number of protons, both of which form the nucleus. The number of electrons, is also almost always the same as the number of protons, these numbers being determinative of the element they comprise. The neutrons have no electric charge. The simplest element is hydrogen with one proton and one electron. Helium has two protons and two electrons. Carbon, nitrogen and oxygen, each of which is so vital to life, have 6, 7, and 8 of each respectively. Uranium has 92 of each. There are 98 naturally occurring elements. Number 98, consisting of 98 one of each, is named Californium, which gives us an idea of where it was discovered. There are 20 other elements created thus far only in the laboratory. The atomic number is the only difference between them. Otherwise, an atom of carbon, for instance, in a stone, a plant, or a human, or other animal, are all precisely alike.

The protons and neutrons are made up by the most fundamental subatomic particles presently known, the ones termed 'quarks.' Physicists assume also that every particle has an antiparticle, some of which have already been discovered, including the antiquark, with a negative instead of positive charge, which sometimes appears in combination with quarks.

The protons have a positive charge, the neutrons have no charge. Hence the nucleus has a positive charge which balances out the negative charge of the electrons. The quarks of which the nuclei are composed are six in number. Quarks differ from one another in their mass and charge. Their names have been whimsically bestowed by physicists who usually talk about them in terms of three pairs: up/down, charm/strange, and top/bottom.

Quarks appear to be true elementary particles; that is, they have no apparent structure of inner parts and cannot be resolved into something smaller. The same is true of electrons. Quarks always seem to occur in combination with other quarks or with antiquarks.

Considering that like charges repel each other, why do not the protons, all of which have positive electric charges, repel each other? It is because the quarks that compose them are all bound to each other by the 'strong force' consisting of gluons, which are even smaller than the protons. The strong force is one of the four fundamental forces recognized today by physicists, the others being the weak force which plays a role in unstable atoms, the electromagnetic force, and gravity. Gravity, which is all around us in our macro-world, that is, of large objects, plays very little role in particle physics.

There are a few other things we should know about atoms before hearing something about entanglement. First, how many atoms are there in the world? There are some estimates, but no one can unequivocally vouch for their correctness. We start with the human body. In a body of about 155 pounds, there are estimated to be 7×10^{27}. That is 7 times ten to the 27th power, meaning the ten is followed by 26 zeros. We must remember that each additional zero in the superscript, meaning the degree of magnitude, increases the entire number before it by ten.

How many atoms are in our planet? The most frequent estimate is 1.33×10^{50}, or 10 plus 49 zeros. One can write them out and get an idea of the number it represents, except that we cannot really wrap our minds around such a quantity.

How many atoms are in the known universe? One educated estimate is 2.4×10^{80}. If you would like a name for that many zeros, forget about trillions or quadrillions etc. The word is vigintillion.

Our next important question is: How long do atoms last? What is their life span? It may be easier and more practical to look to the life-span of one of its key components, the reason being that there are ongoing efforts to determine the life spans of these components experimentally. It has been determined, based on experiments that the protons, part of the nucleus, have a 'half-life' of 10^{33} years, a ten followed by 32 zeros. A half-life means that half of them would survive for that number of years, which, in this case, is a billion X a billion years, perhaps the minimum that should be projected.

By contrast, the life of our solar system is estimated on sound authority to last only another 5 billion years, which is about half of its total projected life span, the first half having already passed. The comparison between the two lifespans, of the solar system and the universe, is like comparing a family of 5 to the entire population of the Western Hemisphere; or comparing a five dollar bill to a billion dollars. But these life expectancies are averages. In measuring the decay of particles, many will have a much shorter life, others much longer. Some will be very short. If the average age of an American resident, for instance, was found to be 75 years, we would still know that there would be some, relatively few, deaths of children, and other young people.

Michio Kaku, in his *Hyperspace*, published in 1995, spoke of methods being used to detect proton decay. The first attempts were in the late 1980s, and six detectors, all very complex and expensive, were in operation in the United States and Japan. It was anticipated that some proton decay, the huge number of them considered, would occur, even in a few years. At the time of the publication of his book not a single proton decay had been detected. The life of the proton was soon recalculated from 10^{29} to 10^{33}. According to an issue of *Symmetry Magazine* on line, published in September 2015, "Protons, whether inside atoms or drifting free in space, appear to be remarkably stable. We've never seen one decay."

Whatever that half-life could be, it is gigantic when compared to many other life expectancies. We have mentioned the estimated 5 billion years remaining life of our solar system. The universe has existed now for about 13.5 billion years, long before the life of our particular solar system. No one that I know of has estimated the remaining life of our universe, but it seems to be understood that ultimately when all of the protons are gone, so, necessarily, will the atoms, and hence, the universe.

What does this mean for life on Earth? Nothing really. Our doom is scheduled to happen by other means even before the actual destruction of our solar system. We cannot expect that life will continue with of all its pleasures and horrors for five billion more years, then suddenly dissolve one morning. René Heller, an astrobiologist, says that within a half billion years the expanding sun will threaten the life of all complex multicellular life on Earth. That's us; and on a cosmologic scale, it's like tomorrow. By 1.75 billion years, he predicts, our world will be hot enough to evaporate the oceans. And that will kill any simple life hanging on to the surface. To us, how long the protons, the atoms, or the universe lives, is irrelevant.

To the human race on this planet it is forever. So we can be as certain as science can make us that there is a vigintillion of atoms in the universe, a billion x a billion atoms in our solar system and that from our perspective they will last forever.

One more item about atoms, surprising perhaps, but not really weird. If these building blocks are so tiny, how is it we weigh so much, even the thin ones of us? We are indeed made up of trillions of these particles, but that is far from being the whole answer. We talked about distance within the atom, in the course of which we imagined a nucleus expanded to the size of a golf ball. How much would the expanded nucleus, the virtual golf ball, weigh? Or, in the language of science, what is the density or the mass? Answer: Plenty. The protons and neutrons are the weightiest, most massive, substances known on Earth. Nearly all the mass of an atom is in its nucleus which occupies less than a trillionth of the volume of the atom. If you could expand it into something the size of a large pea, never mind the golf ball, they would weigh about a billion tons each.

We said that the proton is the most massive, or weightiest on Earth. That is probably true, but not the most in the universe. When stars of a mass, of 4 to 8 times that of our sun, die from using up all their fuel, namely hydrogen, they become neutron stars. Their gravity condenses and squeezes together the atoms and its electrons and protons, cancelling the electric charge. Though they are still 1.5 to 3.5 as massive as our sun, they are condensed into an area about the size of Manhattan. At least one physicists has placed the weight of a teaspoon full of such a star at 5 ½ billion tons. Two others speak of 100 billion tons for the same teaspoon full.

But what is it that makes the behavior and characteristics of the atoms so strange, so weird? Nothing really that has been said so far, despite the eye popping nature of the pertinent figures. So it is about time we said just why so many people, including topflight scientists have said that it is weird. The weirdness consists entirely in their behavior. There are behavioral aspects other than entanglement that fit that category of 'strange' and 'weird,' which we should consider first. It is not only necessary that we be familiar with them but they will lead us inevitably into our subject, namely entanglement.

But first we examine two other samples of possible reincarnation, both involving subjects who were murdered, one a very easy one to follow, the other, more complex. The first was developed by a Doctor, one, Dr. Eli Lasch, and reported by Trutz Hardo, one of the leading figures in this field. The second is by Dr. Ian Stevenson, both of which we will see in the next chapter.

Chapter 4 :

Two Cases of Murder

W e look now at two cases of murder, both involving children with memories of their own murders in a 'previous life.' One is clear cut and easy to grasp, the other, perhaps just as clear, but only after more complex research. We turn to the easier one first. It comes from the work of a physician, Dr. Eli Lasch. His account was translated and published by Trutz Hardo, another of the leaders in this field. It is in his book, *Children Who Have Lived before.* This case has received wide publicity since first publication of the book in the year 2000, but is worth summarizing in this context.

It occurred in the population of the Druse, a population of about 200,000, whose religion is limited to one area. They reside in Israel, mostly on the Golan Heights, and are firm believers in reincarnation. Hardo writes that most cases of past life memories are among people who believe in reincarnation, as the parents in such cultures "tend to listen to what such children are saying, and help them make contact with the people from their past." Sadly, as we shall see in the case following this one, this is not always true. Occasionally the child speaking of such things is severely punished as, for one reason among others, the subject's parents fear the child will leave and return to the 'former parents.' But we concentrate now on the Druse case.

When a child is born in this population, the body is searched for birthmarks. This is because they believe something studied scientifically by Ian Stevenson and about which he found much supporting evidence, namely that such marks may stem from death wounds from a previous life. At about age 3, if the child can distinguish between events from the past and present life, and has mentioned a place of the previous life, the child is taken to that place.

The inquiry is conducted by several respected elders. In the case considered here a three year old boy had on his upper forehead from birth a red mark stretching to the center of his head. The boy had said that a man had killed him with an ax, but that he could not remember the name of the murderer, nor his own name. Fifteen men, including the father and other relatives were appointed to investigate. Dr. Eli Lasch was the only non-Druse in the group. In the first two villages they reached, the boy stated that none of them were his previous home. In the third village he told the group that this was where he lived, whereupon he suddenly remembered his own first and second names, and those of his murderer.

One of the village elders claimed to have known the man whom the boy named, and said that he had disappeared without a trace four years ago and had been declared missing. Such disappearances were not unusual in this war torn area. Going through the village the boy identified his house. An inquisitive crowd was starting to gather.

The boy walked up to a man, called him by name, and asked if that were he. The man acknowledged his name, whereupon the boy replied: "I used to be your neighbor. We had a fight and you killed me with an axe." Dr. Lasch told Trutz Hardo that the man turned white as a sheet. The boy continued "I even know where he buried my body." The man apparently had not disappeared.

It should be mentioned that there are a number of other cases wherein a subject, explaining his memories of death of the prior personality, often considering such personality to be his own previous life, reported having seen his body and surrounding events, such as relatives crying. The substances of the description have often been verified by the attendant witnesses. This has also been reported in numerous 'near death experiences.'

The investigating group, followed by many of the inquisitive bystanders, led by the boy, soon walked off into a nearby field. The identified murderer was asked to come along. The boy stopped in front of a pile of stones, and said that the man had buried him under those stones and, pointing, he said that the ax was buried over there. The stones were removed and the skeleton of an adult man was found. A slit in front of the skull was very visible. With everyone staring at him, the identified man admitted to the crime. They then went to the place where the boy claimed the ax was buried. It was soon found.

In response to the question of Dr. Lasch, the investigators said they would not hand the murderer over to the police, but would themselves decide on the appropriate punishment. What it was we are not told.

Assuming the facts are accurately reported, such an uncomplicated and direct chain of events is potent evidence for survival of individual memories, mostly, but not always, in infants or young children. Memories of events occurring shortly after the death of the prior personality seem to survive even when little or nothing of the rest of the 'time in between' does.

There seems little reason to suspect the word of either Dr. Lasch or of Mr. Hardo. Apart from their own unquestioned reputations, there are too many other similar cases, and this one stands out only for its brevity, directness of investigation and entirely satisfactory ending. First published in 2000, it could easily have been refuted or contradicted were it not accurately reported, but there has not been any such suggestion. We look now at a case more complicated but of equal probity.

*   *   *

The case of Ravi Shankar:

On January 19, 1951, Ashok Kumar, age 6, usually called Munna, was enticed away from his play and murdered with a knife or razor. He had lived with his father, Sri Jageshwar Prasad (Prasad), a barber in Kanauj, India, in a district named Chhipatti. The severed head of the boy and some of his clothes were later found and identified by his father. The murderers seem to have been one, Jawahar, also a barber, and Chaturi, a washerman; the motive being the disposal of Prasad's sole heir, so that Chaturi, a relative of Prasad, might inherit Prasad's property. Someone had seen Munna go off with the two men. Chaturi confessed, but after being charged withdrew his confession. Without witnesses, the charges against the two men were dismissed.

A few years later Prasad heard of a boy, born in in a nearby district in July 1951, six months after the death of Munna. He had described himself as the son of Jageshwar, a barber, in the district of Chhipatti. He also named the murderers, the place of the crime, and some details of the life and death of Munna. The boy's name was **Ravi Shankar** (Ravi). His father was Sri Babu Ram Gupta (Gupta). Between the ages of 2 and 3, the boy often asked his parents for toys he claimed to have had in the house of his "earlier life." Later, when he was almost 6, his school teacher listened to the boy's narration of the murder.

Dr. Stevenson began his investigation of this case in 1963, and in 1964 visited the site and interviewed numerous witnesses. He concluded that, based on a number of circumstances, the two families involved had had only the slightest acquaintance with each other. The mother of Ravi (Ramdulari Gupta), had not known the other family at all, but went, as did many in that area to pay condolences to the parents of Munna, the brutal murder having been widely publicized and the topic of much conversation. She stated also that her son had a linear mark from birth, first noticed by her when Ravi was three to four months old, which seemed similar to a knife wound across the neck. When Ravi began talking he would say that the mark was from his wound from the murder as Munna in his previous life. Ravi's father (Gupta), opposed any discussion about the matter, and beat Ravi severely, apparently fearing that the boy would want to 'return' to Prasad's family.

Gupta, in fact, had insisted that everyone, neighbors included, forget the entire incident. Ravi's schoolteacher in 1956 observed the effects of the beatings by his father. Gupta, in fact, sent Ravi away from the district for a year. Gupta subsequently died before Stevenson's investigation. In addition to fearing his father, Ravi had been fearful of the murderers of Munna. He had trembled with fear and had run away on seeing either of them and told his teacher that he was afraid of all barbers and washermen. Ravi, nonetheless, had sworn vengeance against them.

Stevenson interviewed numerous witnesses, took written statement from five: Mano Rama, (the mother of Munna), Prasad, (Munna's father), and three of his neighbors. Verbal testimony was taken from four: Ravi, Ramdulari Gupta (Ramdulari), mother of Ravi; Uma, older brother of Ravi; Maheswari (older sister of Ravi); Kumar Rathor (neighbor of Ravi); Shiriam Mishra (school teacher of Ravi), plus various letters and documents.

Stevenson listed 26 statements made by Ravi relating to the life of Munna. Each of them was confirmed, or partially so as indicated below, by one or more of the witnesses named above. Other than the basic facts of the murder already set forth, the statements, included the fact that he (as Munna) was eating guavas before he was murdered; that he was enticed by the murderers with an invitation to play a game called Geri. No one could confirm this, but it was said to be a game that Munna often played with Chaturi and Jawahar. Stevenson says the statement is "probably" correct.

Munna had a toy pistol at home, something poor people such as Prasad could not usually afford. Inasmuch as Munna was his only son, he bought one for him. This and the other toys about which Ravi now inquired, a toy elephant, a statuette of Lord Krishna, a ball attached to an electric string, a watch and a ring were still in the home. Ravi also identified his (Munna's) father the first time he saw him. He also showed fear and identified his (Munna's) murderer, Chaturi, the first time he saw him. He also identified his (Munna's) maternal grandmother the first time he saw her and said, correctly that she had come from her home in Kanpur.

He had attended the primary school of Chhipatti, and said that his "wooden slate was in the almirah."

Stevenson set forth in his publication of the case, his reasons for belief that any fraudulent purpose was most unlikely, as it certainly was. It does not seem like a necessary exercise, and totally lacking in any ulterior motive. He dismissed also the possibility of unconscious retention by Ravi of things he had heard, and retained in his subconscious though consciously forgotten, a phenomenon called crytomnesia.

In so doing he took into account the fact that Gupta beat the boy for talking about a prior life and tried to forbid anyone else from speaking of the matter. The mother of Munna suffered a complete breakdown, apparently from grief, and her reaction to news about Ravi was decidedly not joyous. She could not bear to think of Munna with another mother. Further, there was no way for Ravi to have learned at such an early age, three years at the most, the details of toys that he had as Munna, and about which he continually asked.

Ravi's attempt to have the case brought against the murderers was unsuccessful. Stevenson does not know why, but conjectures that the judge would not rely solely on the claim of a boy when the boy recalled facts from an allegedly previous life. Stevenson was in agreement that no case should turn on such testimony.

Stevenson rarely, if ever, claims his reports constitute proof of reincarnation. He does not do so here. The fact that he dwells on no weaknesses in the case, as is his custom, in itself should tell us much. The strength of such cases should be kept in mind as we learn more about atoms in the next chapter and some later ones.

Chapter 5:

WEIRD BEHAVIOR OF THE MIGHTY ATOMS

Einstein said that if quantum mechanics were correct then the world would be crazy. Einstein was right-the world is crazy. Daniel Greenberger

Before we talk about entanglement, there remain some behavioral aspects of these mighty atoms, knowledge of which is a necessary preface to it. Probably the two most often mentioned explanations of one of these behavioral aspects, termed 'superposition,' involve, first, the thought exercise known as 'Schrödinger's cat,' the other known as the 'double slit' experiment. The double slit was first performed by a 28 year old highly inquisitive scholar of many interests over 200 years ago, in 1801.

Superposition is a still mysterious aspect of quantum behavior, namely the dual nature of quantum particles, and sometimes of molecules too large to be called particles. The dual nature is their existence with two or more characteristics which we consider contradictory to each other and are normally, in our macro (large)-world, not seen to coexist. The most common of these is the dual appearance of certain particles with characteristics of both waves and particles. It was first found to be a characteristic of light. The duality, or 'superposition,' can last until observation by a detector forces one characteristic to prevail, the other to disappear. Until then, the wave is usually termed a possibility wave there being so many possibilities that could result from the observation.

What can also seem rather odd, at least to me, is the fact that the thought exercise of Schrödinger's cat has been around so long and is still part of any book and many papers on the wave/particle duality subject. It is doubtful that Erwin Schrödinger, its creator, ever intended anything more serious with it than to demonstrate the absurdity of anything other than atoms or sub-atomic particles being both wave and particle at the same time. The cat parable has been mentioned often enough that we should touch on it here, but perhaps irrelevant to the degree that we should make it short.

Most physicists believe that Schrödinger was demonstrating the impossibility of the superposition in our 'macro' world objects, such as cats or automobiles. In his parable we have a cat in a sealed box, with a deadly poison that can be released only if struck by a radioactive particle from another substance in the box. Because it is a subatomic particle, according to the laws of quantum physics it is, until observed, in a superimposed state. It may be decayed or not decayed. Which means that the cat is in a superimposed state of alive and dead, something one physicist called "patently absurd." But when you open the box and look at it, continues Schrödinger, the cat is either alive or dead; it is one or the other. The big question, and the big point, before you open the box, is what is the cat's actual condition? Alive *and* dead?

Others have emphasized the mathematical and statistically extreme unlikelihood of such superposition affecting the everyday objects in the macro world we live in, a world of large objects, such as cats and cars, pins and needles, elephants and houses. Such a thing, it is said, might by chance occur once in a period of many times the age of the universe. It would be like flipping a coin a trillion times and getting heads every time. Virtually all the atoms would by chance have to take on just the right characteristics.

So let us turn to the second frequently cited example, the one termed the double slit experiment. It proves, reproves, and proves again the reality of superposition and of the wave particle duality. It is one of the most important lynchpins in the structure of quantum physics and some familiarity with it is necessary to all but the most superficial understanding of entanglement.

As already mentioned, as early as 1801, such an experiment was conducted by Thomas Young, the precocious English physician and physicist. At that time there was already some controversy about light and its nature: was it a wave or did it consist of particles. Newton had believed that it consisted of particles. Young was convinced that light was a wave. So he conducted an experiment. He used a light source which he shinned on a screen. On the screen he cut two parallel slits. Behind the wall was a screen with no slits. If light was comprised of particles, he reasoned, the particles would go through one or the other slit and pile up in two places on the screen like bullet holes striking any object. If light were a wave, the two waves would spread out and 'interfere' with each other. This means that the amplitude of the two waves would at various points combine, producing a peak on the screen, and alternately cancel each other out completely, leaving a trough. In short, the stripes of different intensity would signify higher or lower density than undisturbed light. Young found the alternating stripes which convinced him that light was a wave.

But now fast forward to the early 20th century. In December 1900 the German physicist, Max Planck, doing research on what is known as 'black body radiation,' namely radiation from a hot body, published a paper establishing that energy levels are not continuous, but were always a whole number multiple of a very small number, a basic quantum, known thereafter as Planck's constant. All energy, that is, came in packets that were 2, 3, 4 etc. times Planck's constant, never 3 ½ or 6 7/8, or 6.37 etc. This is highly indicative of light as particles.

Albert Einstein, explaining the photoelectric effect which he discovered in in 1905, and for which he won the Nobel Prize in 1922, was convinced that light, as Newton believed, did indeed consist as particles. The particles were soon named photons, a word of Greek derivation meaning 'light.' Particles they may have been, but they were also waves, and in the mid-1920s, the French physicist, Louis de Broglie claimed that not only did the massless photon have wavelike properties, such as interference, but so did bodies with mass, including much larger particles such as electrons.

In the mid-20th century came the results of progress in another field. It included a refinement of equipment that enabled the firing of a single photon and other laboratory devices such as quantum beam splitters and interferometers. This made it possible to merge two or more sources of light to create an interference pattern, which can be measured and analyzed.

Their workings are not important to us; their results are. A single photon was fired at the double slit wall. The physicists expected to be able to observe which slit the photon traversed and whether it showed upon hitting the screen as a particle or a wave. Surprisingly it not only was spread as a wave but showed the same interference pattern. But what did it, the single photon, interfere with? It was this that gave rise to the belief that the photon had been in two places at one time and had interfered with itself. So said and wrote most authorities.

At least one of them however dissented. Brian Clegg, in his *The Quantum Age*, wrote that it would be more accurate to say that the photon, both in Young's experiment and in the modern one, isn't anywhere until it hits the screen and is registered. Up to that point it is a probability wave that encompasses both slits. If, he continues, we put a detector in one of the slits that lets a photon through, but detects its passage, the interference pattern disappears. "We have forced the photon to have a location and there is no opportunity for the probability waves to interfere. However though the photons will indeed pile up behind the slits like particles, an increasing number of particles will eventually take place to make the entirety to look like colliding waves with its alternating dark and light portions.

De Broglie's idea that all particles exhibit wave phenomena was confirmed in the mid-20th century, by experiments with electrons, neutrons and atoms. Each has multiple characteristics that can involve superposition. To mention two of the most frequently used in experiments, there is polarization of photons and spin, up or down, of electrons. Measurement by a detector would force a decision on the wave. A particle would appear. As long as it was not measured it was a wave. Measure it and it was a particle though its characteristics could not be predicted.

But nature is not through with us yet. To make a measurement is to destroy the integrity of the particle. At the same time you determine its location or other attribute it is destroyed.

The scientists themselves have not been altogether consistent in their use of terms. Some refer to the particle as coming into existence (only to be destroyed) when "observed or measured." Others speak of it as measured only, which means by the detectors. I would opt for measured without the ambiguous 'observed.' To this day no one has ever observed an atom, or subatomic particle with the unaided eye, though they can see a wave.

Yet one source left no doubt about what he meant by 'observed.' In a not very helpful illustration he asks if we have ever seen a field goal kicked that went both through the uprights and under it. Granting the absurdity of such a result, he explains that even if that were happening, the observation by so many pairs of eyes, namely those of the spectators, would force a choice upon it. What would result if one half each of the spectators forced different results, he does not say. We have already dealt with the nigh impossibility of such a thing. Our illustrator-scientist makes clear however that he understands 'observed' to include seeing with the unaided eye.

But, say many others, according to quantum mechanics, wave-particle duality and quantum superpositions must also occur for macroscopic objects such as viruses, cells and even larger objects. And that raises an interesting question: how big an object can physicists observe behaving like a wave?

Of course, nobody has seen the quantum superposition of a football and the experiment would be impossibly difficult in any event. But physicists have seen this wave-particle duality for increasingly large molecules such as, in 1985, buckyballs, a shorthand term for Buckminsterfullerene. It is a very common naturally occurring molecule found in soot. It has 60 atoms, and at the time was one of the larger molecules to exhibit wave particle duality.

But progress continues. In 2011, two diamonds, very small, but very definitely part of our macro-world, and perceivable by our unaided vision, were separated by 15 centimeters. Some molecules within, which could not be seen with the unaided eye, were not only superimposed, but were entangled, about which we will shortly see more.

At the University of Vienna in November 2013 a few physicists set a record for quantum superposition by observing wavelike behavior in giant molecules containing over 800 atoms. A beam of these molecules was passed through a series of slits. The molecules then did indeed form an interference pattern at the detector which meant that they were superimposed while passing through the slits.

The team said that this confirmed the "fully coherent quantum delocalization" of single compounds composed of about 5000 protons, neutrons and electrons. The 'delocalization' refers to the electrons becoming part of two or more atoms, which means a bonding of atoms. The team's measurements indicated that this molecule has a wavelength of about 500 femtometers, each femtometer being 10^{-15} of a meter. This is a decimal point followed by 14 zeroes and a 1, which is about four orders of magnitude smaller than the diameter of a single such molecule. This explains the large number of protons, neutrons and electrons. That is a significant step forward for the detection of wave-particle duality and quantum superposition in large, macroscopic, objects.

However, it still leaves open the question of just how big an object can be and still be observed forming a quantum superposition. These molecules are of course tiny but significantly they are within an order of magnitude or so of the smallest viruses. If that is the next step, its significance could hardly be overestimated.

Neither can the other deep mystery, entanglement. That comes next, but only after we hear a little more, in the next chapter, about children who claim to have lived before.

Chapter 6:

TO COMPARE AND CONTRAST: IOWA AND HUNGARY

S igns of memories of another life can occur in no particular pattern and cannot be limited to any particular status, rich or poor, educated or uneducated, and people white, black, brown or yellow. They do seem to be born more often to parents who believe in reincarnation than to those who do not. Those who do believe, even if motivated to play down the extraneous memories for one reason or another are usually less likely to punish the child or to discourage it by dismissing it as nonsense. We will look now at two cases which seen together can sharpen that observation. We start with a case from Iowa.

Romy Crees, the daughter of Barry and Bonnie Crees, was born in Des Moines, Iowa in 1977. The parents were both devout Catholics. When Romy began to talk, they were thoroughly bewildered and at a complete loss to understand certain things said by her in addition to the usual childhood talk. Their reaction was understandable as it involved claims of a previous life as a boy named **Joe Williams.**

Among the details she mentioned was having grown up in a red brick house in Charles City, about 140 miles from Des Moines. She also claimed that, as Joe, she had married, and that her (his) wife was named Sheila; that they had three children, and that both Joe and Sheila had been killed in a motorcycle accident which she described in detail. She claimed to be afraid now of motorcycles. She also claimed that, as Joe, she started a fire in the home of her (his) mother, whose first name was Louise; that the mother burned her hand throwing water on the flames, and that she had a pain in the right leg. Romy claimed she hadn't seen her former mother, Louise, and often asked to be taken to Charles City to assure Louise that she was OK.

Romy's parents were convinced that this was all a fantasy, weird though it was, but the detailed description of Joe's life and death forced them to have second thoughts. As it happened, Hemendra Banerjee, a professional investigator specializing in such cases, came with his wife in late 1981 to Des Moines, accompanied by two journalists from a Swedish magazine, *Allers.* The couple willingly submitted to interviews of themselves and Romy. The entire group later set out for Charles City to see if the claims could be verified.

Romy was visibly excited during the drive. As they drew near Charles City she made a number of significant statements. She exclaimed that they had to buy flowers; that Mother Williams loved blue flowers; and that they could not go through the front door of her home, but would have to go around the corner to the middle door. With the assistance of the telephone directory they found, and soon pulled up in front of a white bungalow. It was not a red brick house as Romy had described, but on the walkway was a sign stating "Please use the backdoor." Romy jumped out of the car, practically pulling Banerjee up the walk with her.

An elderly woman answered the knock on the middle door. She was using metal crutches and wore a tight bandage around her right leg. In answer to questions she affirmed that she was Mrs. Louise Williams; that she once had a son named Joe, but stated that she had a doctor's appointment and would talk to the group again later. About an hour later Romy, her father and the journalists did return and were welcomed inside.

Louise was startled, and touched, by the gift of blue flowers, explaining that her son's last gift to her had been a bouquet of blue flowers. Romy's father, Barry Crees, narrated the details that he had heard from his daughter about Joe's life, Mrs. Williams asked in astonishment, "Where did this girl get all this information? I don't know you or anyone else in Des Moines."

About the red brick home, she explained that she and Joe had lived in a red brick house, but that it had been destroyed by a tornado that damaged much of Charles City about ten years ago. She also explained that she and Joe helped to build the present house and that it was Joe who insisted that the front door be shut during the winter.

When Mrs. Williams stood up to get something in the next room, Romy rushed to follow her. They returned hand in hand, Romy trying to help Mrs. Williams walk. The elderly lady was carrying a framed photograph of Joe and his family taken the Christmas before he and Sheila were killed. The surprise still registered on her face. "She recognized them. She recognized them," Mrs. Williams repeated. She could also confirm much of what Romy had narrated earlier to her parents, the Crees. That included his marriage to Sheila, the three children, names of other relatives, the fire in the house which had burned her hand, and the precise details of the motorcycle accident which occurred two years before Romy was born.

All of these circumstances notwithstanding, neither the Crees nor Mrs. Williams was prepared to accept any notion of reincarnation. "I don't know how to explain it," said Mrs. Crees, "but I do know my daughter isn't lying.

There is nothing too unusual about such apparently contradictory reactions. We are creatures of our emotions. For the Crees, particularly, rejection of reincarnation may be bottomed on religious belief. To disavow any suspicion that rejection of the obvious does not impugn the honesty of her daughter is equally necessary. That may have an equally strong root, based on filial love.

This and the following case, especially when considered together, well illustrate that in addition to our genes and epigenetics, and even to 'reincarnation,' which I prefer to call a 'third level of inheritance,' we are also products of our environment and the culture into which we are born.

*   *   *

Gedeon Haich

We have seen the case of a parental couple who would not, and could not accept the possibility of reincarnation, even in the face of the memories of their young daughter which obviously came from a boy in a distant city. Even the clear evidence of reincarnation, or something akin to it, was beyond them, and they first reacted with a certain disdain for her talk of such things. I mentioned the culture in which they lived as being a strong factor.

I would like now to show an example of a mother who lived in a completely different culture, namely Hungary, and her own open mindedness to what we like to call paranormal, and the difference in attitude toward her offspring from that of the Crees.

The parent is Elizabeth Haich, mother of the subject of this case, **Gedeon Haich**. She lived in Budapest, Hungary. Why would Hungary be so different from Iowa? Hungary is in Europe, perhaps not what would be called Western Europe, but at least right next door. But whatever the similarities or differences, the belief in reincarnation is comparatively huge.

Gedeon was born in 1921. The most definitive statistical figures of interest are the results of a 2011 survey taken by a Paris based global research company, *Ipsos*, for Reuters News. Percentages were sought for a number of beliefs, including the existence of a God, or gods (45 %) or Devine Being (6%).

Turning closer to our area of interest, Just over half of global citizens (51%) say they believe in some form of afterlife: one quarter (23%) believe in an afterlife "but not specifically in a heaven or hell", two in ten (19%) believe "you go to heaven or hell", and 2% believe in "heaven but not hell". Alternatively, one quarter (23%) say "you simply cease to exist" whereas another quarter (26%) say they "don't know what happens".

Homing in on our area, another 7% believe "you are ultimately reincarnated." Hence reincarnation is but one category of an after-life, and a small set of the whole, but the one of immediate interest to us.

Where are most of that group to be found? "Of those who believe "you are ultimately reincarnated", "Hungary (13%) is at the top followed by citizens from Brazil (12%), Mexico (11%), Japan (10%), Argentina (9%) and Australia (9%)". The division into such categories may explain the difference between these figures and those reported later in Chapter 11. When questioning individuals, so much depends on what and how the questions are asked. Do you believe in reincarnation is quite different from do you believed in life after death. Heaven and hell refers to life after death, but not necessarily, in fact at odds with, reincarnation.

In any event, the culture, the atmosphere, the feel of the country must have contributed to some uncertain degree in the life of Elisabeth Haich. She wrote many books, including one that is entitled *Einweihung* (Dedication). It was said about this volume. "The reader will find in this book an astounding richness of thought, experience and learning, encompassing the entire area of the mysterious, and an overview of the most impressive of the laws of karma and reincarnation. In this book in which biography, novel and mystical teaching is mixed, Elizabeth Haich succeeds in painting a tension filled picture which stimulated the profound truth of earlier fantasies of humanity."

Another author described it as "A book about unique wisdom which lay long hidden, and now discovered, the eternal body of laws governing the way of the spirit is now open. A mystical biographical novel which can answer the spiritual questions of our time, and make possible a new view of life."

One of the readers of that book was Dr. Ian Stevenson. We pick up the story from there.

Gedeon Haich, only child of his parents Subo and Elizabeth, was born in Budapest, Hungary on March 7th 1921. His parents divorced when he was about 3 ½ years old, and Gedeon lived with his mother and her sister who shared a joint household. The sister, Gedeon's aunt had a daughter about Gedeon's age and the cousins became companions. When Gedeon was seven his father gained legal custody, though the boy spent summers with his mother and had frequent contacts with her.

Elizabeth kept a diary during her son's childhood until he went to live with his father. Stevenson had two long sessions with her in February and May 1964 after reading her published *Einweihung*.

He did not speak of a previous life between 2 and 4 years, as do most children who remember previous lives. However sometime between 4 and 5 when he and his cousin were drawing together, Elizabeth noted something peculiar.

Though the cousin always gave her human figures rose colored skin, Gedeon always colored his dark brown. With Elizabeth's suggestion that he not make his human figures so dark, Gedeon said nothing, but continued with his brown coloration. Stevenson met Gedeon for a long interview in November 1972 and describes his appearance as typically European: blue irises, light brown straight hair with no suggestion of short curls characteristic of Africans, nor were his lips as thick as Africans. Gedeon was 51 at the time, but claimed that he still had some memories, though the images were not clear.

Shortly after the episode with the skin coloration, Elizabeth found that her son resisted and screamed when asked to join other family members when swimming in a lake near their summer home.

The biggest surprise however came when he was between 6 and 7 years old. Before he went to live with his father, he asked his mother whether he might have lived before he became her son. To his bewildered mother's questions he said that he remembered being in a different country where he had a wife and children. With pencil and paper he made a drawing that depicted a small hut with smoke coming out of a vent from the side, something that was not to be found in Hungary. In a walkway leading from the house he drew a naked woman with long sagging breasts. There was a nearby body of water with waves, and in a different picture he showed palm trees.

Gedeon explained to his mother that he and his family lived in huts like he drew and that each member made a boat for himself by carving out the trunk of a tree. There was a large river nearby, but they could not go into it as deeply as the lake by the summer home as "A kind of monster lived in the water." That was why he resisted so against going into the lake. He was afraid, he explained, that there was something that could bite his legs, and that he was still afraid though he knows there is nothing dangerous in the lake.

He told his mother how he became convinced of life after death from watching a beetle lie perfectly still for a half hour then turn over and start moving Then, more substantively, he added that in the mornings, his eyes not yet open, he immediately feels that he must jump out of bed and go hunting to find food for his wife and children. It was, he said, only upon opening his eyes and looking around that he remembered he was a "small boy and your son."

He reminded his mother that she did not want him to row a boat that the family bought until he learned to row, and how he assured her that he could. She challenged him to show how he rowed, and he surprised her by rowing expertly, though due to his small size he could use only one paddle. Where he lived, he said, he could do anything with a "treeboat." The trees were different there from the ones where they now were. He drew another sketch that showed palm trees, a man hunting with bow and arrow, and a hat, surprisingly European in style, resting on the ground. There was also a curved object in the air.

Stevenson asked him why he drew his wife with such "long hanging, ugly breasts," to which Gedeon replied "Because that is the way they were. And they were not ugly. She was very beautiful."

Stevenson asked about the last thing he remembered. Gedeon replied that he was hunting and threw his spear at a tiger. It hit the beast but didn't kill him. "The tiger jumped on me with the spear still in him. I don't remember what happened then."

At one point Elizabeth asked Gedeon about the curved object in the air. He said that he had made it himself. It was a weapon that "one threw and that returned back by itself."

Otherwise, Gedeon talked little about his prior life. At age 13 a neighbor demanded that he come down from a tree he had climbed, which she estimated to be at least 65 feet high. She noted that he came down skillfully "like a small monkey." Asked by his mother why he climbed such a high tree, he gave his reasons, to build a nest for cooking corn, the food was better and the view was better. Ordered by his mother next time to build his nest on the ground, he somewhat angrily replied "I would like to know who looked out for me when I was in the jungle and climbed even higher than this one to watch for animals. Where were you then?"

He came home from school once quite angry that a clergyman said that we live only once. Gedeon said that he knew we lived many times, but that among grownups it was better to be quiet.

At age 15 his mother bought him a drum. She was astonished at his skill in playing the most complicated rhythms. She thought he seemed to be in a kind of ecstasy with tears in his eyes. Her son replied, "Do you see Mama that is the way we could send signals and messages to each other over great distances."

His mother wrote in her book, *Einweihung*, that Gedeon had never been able to see a book about Africa and that she knew every step he took and how he had been occupied, and he had never been in a motion picture theater. She said that the family had no social or commercial connections with Africa or India. In a 1964 interview, she emphasized that he could not have acquired knowledge of tropical life by normal means.

In later years Gedeon learned and practiced, and ultimately taught yoga as his life's work. Stevenson met with him again in 1972. He learned that In World War II Gedeon joined the Hungarian Air Force, was shot down and wounded, but survived intact. Despite his interest in yoga, he showed none about eastern religions nor about India. But he had a strong interest in Africa though he had never been there.

After considerable analysis, Stevenson determined that his prior life was most probably in tropical Africa or south Asia. He wasted little ink on analysis of the validity of the subject's belief in a prior life. He commented only that his sketches as a young child and unusual behavior cannot be influenced by genes, by epigenetics or by the environment, separately or together. Obviously not, hence only reincarnation, or, as I prefer to call it, a third layer of inheritance is left.

That would be all that is left to explain the intrusive memories, but it is most likely the difference in culture that accounts for the behavioral differences of the parents of Romy Crees from the mother of Gedeon Haich.

Stevenson's belief in the high probability, at least, of reincarnation here is attested to by his lack of highlighting any weak spots in the case. He almost never expresses a high level of belief, though many times such level, in spite of himself, exudes from his comments. This is such a case.

*   *   *

It would be apropos at this juncture to give, at least in sketchy form, an explanation of 'epigenetics,' a term previously used and will necessarily continue to be. This epigenetic change can be accomplished through any part of about 98 % of the entire genome, only @ 2% being involved in the better understood genetic inheritance. Epigenesis involves recognition of the fact that, contrary to previous belief, there can be, under certain circumstances, inheritance of traits acquired by the parents during their lifetimes. It was previously believed that 'inheritance of acquired characteristics' was not possible. Change occurred, it was believed, through mutations in the genes involved in procreation.

Heritable changes through epigenetics is not necessarily permanent and may appear only in the first or the first several generations. It can however, through 'canalization,' cause mutated genes to permanently affect the change begun by epigenetic pathways.

Chapter 7 :

ENTANGLEMENT

W e come, at last, to entanglement. As previously outlined it refers to a relationship between two or more atoms, or subatomic particles. These subatomic particles include chiefly electrons, which are negatively charged, existing in clouds around the nucleus of the atom; protons, a positively charged particle in the nucleus; or photons, particles of light. It has also been found to apply to some molecules. Molecules can consist of unions of from two or three, to many hundreds, sometimes thousands of atoms. They are not necessarily 'entangled' with each other in the sense the term is used here.

As explained in Chapter 1, entangled particles have characteristics complementary to each other. Spin is a characteristic of electrons. If one of two entangled electrons spins right, the other spins left. Polarization, which controls spin in photons can be located at an infinite number of spaces on the photon, but two points on the surface are the most that can be measured.

Entanglement once formed will, for our purposes, last forever. Further the relationship continues no matter how far any subsequent separation between them may be, even to the opposite ends of the universe, if there is such a thing.

The change in one particle means a change in the other so as to keep the complementary relationship intact. The change is instantaneous, each changing simultaneously. Edwin Schrödinger, one of the major figures in this field, has said that "entanglement is not *one* but rather *the* characteristic trait of quantum mechanics."

Even today, almost all of the articles and books that address the subject of simultaneity of change in entangled atoms or particles, no matter the distance between them, insist that it does not violate Einstein's 186,000 miles per second speed limit. Their reasons: first, no classical information, meaning a language humans can understand, can be transmitted by entanglement. Secondly, they say, the required examination, that is measurement or observation, would destroy the atom or particle.

It is worth noting that Einstein, who obviously knew all about relativity and the universal maximum speed, which is that of light in a vacuum, never saw it that way. He also knew about the subject of entanglement and its limitations. Yet it never occurred to him to claim that entanglement did not contradict his relativity theory nor any portion of it. Nor has anyone, as far as I can find, ever specifically pointed to any part of his papers on relativity that claimed the maximum speed applied only to information that humans understood, or only to means of transmissions that remained intact after delivery of the message. On the contrary, Einstein's fierce determination to disprove various aspects of quantum physics, including entanglement, seemed driven by defense of conclusions articulated in his special theory of relativity, particularly his pronounced 186, 000 mps universal speed limit.

His chief opponent, a leading champion of quantum physics, was Neils Bohr. Bohr vigorously defended the far stricter idea of quantum mechanics, known as the Copenhagen interpretation. The two men often argued passionately about the subject, especially at the Solvay Conferences of 1927 and 1930; neither ever conceded defeat. It was not until 9 years after Einstein's death, in 1955, that such a means was proposed. His death in 1955 spared him having to witness the ultimate triumph of quantum over 'classical' physics.

The proof was ironically precipitated by a paper published by Einstein himself and two colleagues, Podolsky and Rosen (EPR) in 1935. In this paper, it was stated that 'locality,' the widely accepted idea that one object could not affect any other without some means of contact between them, whether physically or through some force or wave, was violated by the claims of quantum physicists.

Isaac Newton himself had referred to such a possibility as nonsense. It does indeed contradict 'common sense.' EPR claimed hence that some 'unknown variables' inherent in the particles or atoms, before their separation, must be involved, hence the quantum theory must be incomplete. The paper enunciated the proposition that locality was obviously a requirement for one object to affect another.

Time and experiments tended to disprove the unknown variables idea, and another physicist, John Stewart Bell proved that both quantum mechanics on the one hand, and realism and locality on the other could not both be right. Bell further published a mathematical theorem claiming that certain inequalities of results would be expected if Quantum was wrong, but would dictate that Quantum was right if the inequalities were violated. But at the time there was no known way to test such conflicting claims.

There have been numerous theoretical and experimental developments since Einstein and his colleagues published their original EPR paper. Most physicists today, on the strength of the results of myriads of experiments, regard the so-called "paradox" of the paper more as an illustration of how quantum mechanics violates classical physics, rather than as evidence that quantum theory itself is fundamentally flawed, as Einstein had originally intended.

The blueprint for the first experiment was the paper by Bell, in 1964, concerning what he called the EPR paradox. He firmly agreed that Einstein and EPR were correct. He attacked the idea of entanglement, and set out his potential theorem to determine whether quantum physics on the one hand or reason and locality on the other were correct. They could not both be right. The proposed solution was based to a large degree on statistics.

As David Kaiser, an MIT theoretical physicist put it: "No one would expect the roll of the dice to be the same across those two separate crap tables more often than chance . . . Bell's theorem put a mathematical limit on how often these measurements could lineup and agree if the particles were independent." These were the 'inequalities,' and the basis for Bell's claim that if his inequalities could be violated in such tests, that would be evidence of the correctness of quantum theory and against Einstein's claim of local realism.

Bell's theorem assumed a supplemental hidden variable structure carrying information from the beginning of separation. But his theorem clearly showed that no hidden variables could produce all of the predictions of quantum theory, particularly the ones related to entanglement, and that quantum could not be supplemented with them. The violation of the inequalities he specified, constituted evidence of non-locality, and that what happened to one particle does indeed affect instantaneously a second particle, no matter how distant.

The result was a powerful weapon in the hands of the advocates of quantum theory. But Bell was relying on a special case and it did not in itself prove the falsity of the locality assumption as Bell claimed. It did not prove, as previously noted, that both views could not still be valid. That would have to await proof that Bell's assumption was correct.

In a series of experiments conducted primarily by French physicist Alain Aspect between the mid-1960s and early 1980s, proof came. His experiments were preceded by tireless and extended preparations during which he took full advantage of the increased technical knowledge developed during the decades following Bell's work. The evidence in favor of quantum theory and non-locality was overwhelming and practically definitive. The climax came with the proof of a photon that could still send a signal to another and that reaction of the other was instantaneous, despite the induced inability of the photons to signal each other by any normal means of classical physics.

*   *   *

There are other significant matters that must be mentioned.

What does 'instantaneous' or 'simultaneous' mean? In our everyday speech it means 'at the same time.' Which in turn means whatever we want it to. If two people happen to meet at about a certain time at a certain place, and arrived within a few seconds of each other, they might agree that they arrived simultaneously. When quantum physicists work with that word and find that a signal from one particle affects its entangled particle forty miles away, even a millionth of a second would be highly significant. We are, after all trying to determine if the effect on the distant particle was faster than the speed of light, 186,000 miles per second. We are also trying to determine if the effect was truly simultaneous.

True simultaneity has never been shown, and probably never can be. What has been shown is that the effect on the distant particle is thousands of times faster than the speed of light. How close scientists can come to exact simultaneity largely depends on the degree of sophistication of the equipment used.

In the year 2013 a group of Chinese physicists from the University of Shanghai, conducted an experiment that has established the lower limit of the speed involved in interaction between two entangled photons 15.3kilometers (9.5 miles) apart. The tests were repeated over a period of 12 hours in order to establish sufficient data. That lower limit of speed was determined to be 10,000 times the speed of light. That was not the actual speed of interaction, merely the lower limit, as best as could be determined from the available technology.

That is obviously very fast, but not as fast as simultaneity. For communication between points on planet Earth the difference might in most cases be negligible. If at some future time it becomes important to contact people on a distant planet it might not be. Our galaxy is 120,000 light years across. To send a message half way across at light speed would require 60,000 years, and another such time to receive a response. At 10,000 times the speed of light it would still require 6 years there, and 6 years back. Very fast, but obviously significantly different from instantaneous.

*   *   *

What is the longest distance of separation between two entangled particles that has been tested? In 2013 there was a successful transmission between two islands in the Canaries, La Palma to Tenerife, a distance of 144 kilometers (90 miles). The record still stands.

Most experiments with particles have involved two atoms. Three have been entangled on recent occasions. Most recently it was reported by Vladan Vuletic of the MIT Department of Physics, that a single photon entangled 3,000 atoms. Said Vuletic: "It build up correlations that you didn't have before ... We have basically opened up a new class of entangled states we can make, but there are many more new classes to be explore."

New classes have already been opened. One of them involves 'squeezed light,' often called a Bose-Einstein condensate. It and its resulting superconductivity were the subjects of an article in March

2015 by Morgan Mitchell, an expert in the field with the Catalan Institution for Research and Advanced Studies. He spoke of the "striking macroscopic effects" of light and superconductivity, which allows high speed trains to levitate, that is, to ride above the rails, rather than on them.

Mitchell's reaction to this phenomenon are expressed in a later article: "I am continually amazed by quantum mechanics. When the theoretical predictions came out saying there should be a sea of entangled particles in a squeezed state, I was floored. I knew we had to do an experiment to see this up close." Now, he continued, scientists for the first time have been able to directly and experimentally confirm this link. There is however no estimate of the number of photons involved

What does this mean for us and our area of focus? It is circumstantial, but not yet proof that the atoms stored in the brain, though obviously close, are entangled. It was a finding that "In agreement with theoretical predictions, any two photons near each other are entangled." These scientists were dealing with photons as they are the carriers of light. This does not necessarily compel the conclusion that this applies to any atoms that are close.

While we are not told the approximate number of photons involved in the squeezed light, it undoubtedly does not approach the huge numbers of the atoms that comprise the human brain. Any estimate as to how many atoms are involved in that portion of the brain involved formation of, and in storage of long term memories has been extremely difficult to find. We do have an estimate that the entire brain has 1.4×10 to the 26th power, a 10 with 25 more 0s added, all added to by .4 of that number.. If the number of atoms in the motor cortical circuits is not in the trillions it certainly involves many, many billions.

Perhaps the most crucial matter, necessarily answered today only with conjecture, is the mystery of why does entanglement exist. Scientists and technicians are today still doing great things with it. They have or have worked on such things as 'unbreakable codes,' high speed trains, super computers and enough others, such as the internet, that have justified an estimate of one third of our economy. We will turn to the possibility of more fundamental things in due time.

First however we will turn temporarily from hard science to matters dealing more with faith or spirituality, if for no other reason than that it is a part of the entire picture, one believed in by a large portion of the world's population, and one that like the use of science, has resulted in significant proof or likelihood of it. The subject is called regression therapy, and begins often with hypnosis.

Chapter 8:

REGRESSIVE THERAPY AND "THE SEARCH FOR GRACE"

An essay on the subject of regressive therapy may seem out of place in a subject relying as I do on conclusions, sometimes conjectures, based on 'hard' science. But like it or not, and many do not, regression therapy is sometimes part of the total picture. Whatever one's ultimate conclusion, it is worthy of consideration. There appears little reason to reject examination of it, based only upon the wholesale and out of hand dismissal of it by much of the psychiatric community. Many also reject any thought of reincarnation or anything that seems close to it.

Regressive therapy may not involve reincarnation at all. It involves bringing the subject, usually a patient undergoing psychiatric treatment, through hypnosis, to an earlier time in his or her life for the purpose of discovering the cause of some mental or emotional aberration. When the procedure is unsuccessful, some caregivers will extend the hypnosis or use other means to induce a return to a 'prior life.'

The entire subject and all of its branches meet with similar disdain by most psychiatrists for reasons that have been spelled out by author Larry Holcombe in another context and on a different subject. He wrote in a 2015 volume that many in the scientific community reject what is radically new as they are afraid of it. They fear that hitherto "unknown laws of physics will invalidate all they have studied, all the papers and theses they have written, and all they have taught for so many years." Unfortunately there is an historical basis and reality to this thought. It is many of those on the outside of true scientific subjects, Holcombe continued, that "reject scientific examination or even rational discussion of the phenomenon that is so puzzling to those who have made an effort to study the subject."

Psychiatrist are generally taught that all indications of reincarnation or similar phenomenon, result from *crytomnesia*, namely, mistaking for one's own and real experiences, earlier thoughts, or things, only heard or read; *confabulation*, meaning using false or contrived explanations for details that one has forgotten, whether genuinely believed or not; and suggestions by the psychiatrist or other caregiver; or hallucinations etc.

Out of thousands of reported cases, some few are cherry picked for their blatant invalidity and exhibited as similar to all the rest. They are thus taught that reincarnation, either entirely, or at least partially, developed through regression therapy, is for charlatans and/or the misguided. They select particular reported cases for points of real or imagined invalidities and hold them out as representative of the entire body of thousands of cases. I believe, on the contrary, that each individual is entitled to see a broader picture of the field of actual cases. No matter how many invalid cases may be found, if even one appears impregnable to critical analysis, it is significant and invites further inquiry. To quote Max Planck, "Revolutionary scientific advances spend a long time in the heresy box."

We should bear in mind when reading about regressive therapy that its purpose is not to find universal rules for broad application, or to educate the profession or the public at large. That is the purpose of the studies of Ian Stevenson and others mentioned in this essay and in my book. The purpose of regressive therapy is to heal, and allay or reduce the anguish or phobias that prevent the patient from leading a fuller and more productive life. It is the results of the particular case that matters.

One of the most respected and renowned in the field of regressive therapy is the German Scholar Trutz Hardo. He has pointed out the importance of this other aspect of this subject. The goal of regressive therapy, or any therapeutic procedure he maintains, is, or should be, the amelioration of problems that are interfering with the wellbeing of the patient.

Often the symptoms mask the true problem, which, because of its very painful nature is buried deep within the psyche and must be confronted if the treatment is to be effective. By its very nature, the fact that the patient has such thoughts buried within is paramount. Whether the memories are based on fact or imagined through crytomnesia, or confabulation, or leading by suggestion is secondary. To the patient they are just as real.

Trutz Hardo was born in 1939 in Eisenach, which, after World War II, was to become part of East Germany. With Trutz and his four brothers, his father moved to West Germany. In 1960 he earned his first baccalaureate and then pursued studies in history and philology at universities in Berlin and Munich. In 1966 he earned his first degree in Berlin.

In 1982 he co-founded the publishing house of Die Silburschnur located now in Güllesheim on the Rhine. It specializes in books heightening spiritual awareness and what he terms the "Mastery of Life." As publisher and author he has made radio and television appearances, has written several books himself and published cassettes, CDs, and various video tapes. Die Silburschnur has produced more than 180 publications.

He asserts that his study of reincarnation and past-life regressions has shown that problems including phobias and allergies, involving such symptoms as asthma, hay fever, claustrophobia, and impotence can be traced back to root causes, a procedure that often remedies or lessens them.

Trutz Hardo explains that in most such cases the client is taken into a state of waking sleep to youth, childhood, infancy, or, if necessary and appropriate, to a past life. A case in which, through a series of circumstances, the validity of the patient's narrative about past life was confirmed, will shortly be summarized here. We will also see Hardo's analysis of it.

Professionals other than Hardo have practiced and studied this subject. Psychiatrist Professor Adrian Finkelstein, for one, widely known through books and television appearances has stated, the "extraordinary finding" of regressive therapy has changed significantly "the way I now look at illness, the way I look at current existence. It has changed the way I plan to understand my patients from now on and people in general. It opens up horizons in terms of diagnosis and therapy."

Professor Brian Weiss is the author three books on regression therapy. He has written that he "realized that past-life therapy offered a rapid method of treating psychiatric symptoms which had previously taken many months or years of costly therapy to alleviate." He stated that his use of this therapy has produced "excellent results." He further wrote that he now uses past-life regression "with roughly 40% of his patients, while the remaining 60% will benefit from and be adequately treated by traditional approaches. For those 40% however regression to previous lifetimes is the key to a cure."

It is emphasized by Hardo that regression therapy holds the important advantage that neither therapist nor client need believe in the reality of previous lifetimes. Many therapists who have been successfully practicing regression therapy did not originally give the slightest credence to past lives and reincarnation, explaining to their clients who shared this disbelief – that there would be, for instance, symbolic pictures and scenes from other eras arising from their subconscious which could provide important clues for the process of healing.

Hence by their very nature, there is seldom any serious effort for the psychiatrist or other caregiver to play the role of investigator or detective. Often the hidden, yet painful memories are not based on actual experience, and 'exposure' of that fact by triumphant debunkers is, to say the least, irrelevant.

But sometimes the opportunity to verify the truth of the memories, becomes palpable, or sometimes, even falls into one's lap. Such a case is reported by Trutz Hardo in his *30 Most Convincing Cases of Reincarnation"* He Terms it "The Search for Grace."

*　　*　　*

In 1987 a lady in her thirties introduced herself by telephone to Dr. Bruce Goldberg, then practicing regressive therapy in Baltimore. Her name was Ivy and she had heard Dr. Goldberg twice on television. Her avowed reason for coming to him was to find out why she had such a terrible and destructive relationship with a friend she called John, and why she still remained with him.

John, she said, was egoistic, insincere, and unpredictable, and had abused her physically and psychologically. Three times he had almost killed her, but she could not let him go. She was also drawn to another man, named Dave, who seemed the opposite of John. She described him as polite, loving, and trustworthy, though he had not much experience with women.

In addition she was plagued by nightmares in which a man repeatedly murdered her. She had the feeling that it was John, though he looked different and wore different clothes. After 45 sessions she seemed healed of her relationship with John and was having a fulfilling relationship with Dave. Other problems that came up during therapy were also resolved.

But one of her most worrisome phobias, her severe difficulty in swallowing, was not resolved. She would allow no one to touch her throat, not even Dave, nor could she wear anything restricting around her neck. Ivy and the Doctor agreed that this last issue must be probed. They entered into regression therapy and Dr. Goldberg recorded the session and kept notes. She soon saw herself back in the year 1925. She was 31 and her name was Grace Doze. In 1925 Ivy had not yet been born; she could have been born no earlier than 1947.

In the earlier life she was having an argument with her husband, named Chester. They had a one year old son named Cliff. Chester was employed by General Electric. Cliff often stayed with Grace's mother in the same town. Chester often accused Grace of having affairs with other men. Dr. Goldberg's question as to whether the accusation was true she answered in the affirmative. She called her husband, Chester, an idiot, who could not make her happy. They sometimes fought with each other, but she felt strong enough to handle the situation.

Continuing with her recollections of a prior life, the following year they moved together to a flat in Buffalo. She still found her husband, Chester, boring, and often went out alone in the evenings, hitch hiking to her various destinations. She found that being picked up was exciting, as were the wild parties she attended during those times of prohibition. She described a heated argument with her husband on April 19, 1927, in the course of which she injured his arm with a pair of scissors.

She then saw herself two weeks later with a man called Jack. He had apparently fallen in love with her and wanted to stay in Buffalo to be with her. She decided to leave her husband and rented a room in a hotel. She never fought with Jack despite his jealous streak. Chester found her in a hotel and tried to persuade her to return to him. She escaped from him and rented another room. Jack began taking her to a swimming pool where she often swam every week.

On May 17th, 1927 Jack picked her up from the pool. He had obviously been out drinking. He suggested they move somewhere else. Grace insisted that she wanted to take her son Cliff with them, to which Jack vehemently objected. He made disparaging remarks about her clothes and a pair of her shoes with red heels. Becoming angrier he told her that the men in the pub described her as a whore. She denied all accusations. She accused him of being drunk. He called her a tart. He thereupon stopped the car, hit her, stabbed her with a knife, and finally strangled her to death.

There are many instances from narratives of persons who recalled prior lives, who also recall events occurring immediately after his or her death. These are similar to 'near death experiencers,' where the living subject near death is able to see everything from a bird's eye perspective, and without feeling pain. The content of many such reported observations have often been independently verified. Dr. Goldberg, immediately after the description of Grace's death, led Ivy, his patient, into this state of being.

She stated that Jack had thrown her body into the Ellicot Creek. After this regression therapy Ivy lost her neck phobia. Asked if either Chester or her son Cliff had been reborn in her present life, she answered in the negative. She said however that she saw Jack as her earlier friend John.

Dr. Goldberg did not consider it his duty to verify material obtained during regression therapy: "I don't concern myself with names, dates or places, since these have no therapeutic value."

But three years later, going through his notes, he noticed that Ivy had used her full name, Grace Doze, the address in Buffalo where she lived, and the date of her murder, May 17, 1927. Now, as has often been said, the plot thickens. Neither Dr. Goldberg nor Ivy had ever been too Buffalo. He wrote to newspapers in that city to ask whether in the third week in May 1927, there had been a news report of the murder of a woman named Grace Doze.

The results were quite interesting. Three Buffalo newspapers reported a murder that each called mysterious, and which was not solved at the time. Copies of the papers, preserved on microfilm were sent to the doctor. In them were reports of discovery of the body of a Mrs. Grace Doze found in the Ellicott creek. The autopsy showed knife wounds on the body and marks of strangulation on the neck. Her shoes had red heels. For days the papers reported more details. All of the names mentioned jibed with what Ivy had reported under hypnosis, except for two, which, for our purposes will be most important.

The newspapers reported the deceased to be 30 years old; and they stated that the son's name was Chester, like his father. Except for those two, the other details were accurate: the surname, the husband's name, the mother's surname, the street names, the names of the hotels in which she had stayed, her visit to the swimming pool, the name of a friend, Mary, who had accompanied her to the pool on some occasions, and disputes with her husband. The police had questioned the husband, Chester, and held him temporarily as a suspect. We can assume that with no murder weapon, there was insufficient evidence to hold him.

CBS television decided to make a film about the case and the personal history of the principals were thoroughly researched by its investigators. Inveterate skeptics and debunkers often seem to thrive on matching details used by accounts of subjects claiming memories of another life on the one hand, and details reported in history books and other contemporary documents, on the other hand, which 'proved' the subject's details to be erroneous. The errors in the subjects' accounts has often been trumpeted as proof that they were based on contrived or imagined materials and were not true personal experiences. How Ivy, living in Baltimore, Maryland, could have obtained a film of an article in a Buffalo, New York newspaper now over 60 years old, about Grace as a means of fooling anyone, is mind numbing enough.

But in this case, the investigators found that it was the newspapers that had it wrong. It was Ivy, under hypnosis who got it right, both the age of Grace and the name of Grace's son. Had she copied the details from the old newspaper, as Trutz Hardo makes abundantly clear, she too would have given the wrong age of 30, instead of 25 and the name of Grace's son as Chester rather than Clifford, usually shortened to Cliff.

So how do the skeptics explain this set of circumstances? It would be interesting to hear. The naysayers notwithstanding, there apparently can be, it seems, some substance to regressive therapy after all.

Chapter 9 :

THE ETERNAL ATOMS

How do the atoms from one person's organs of memory, come into the body of a growing fetus, or sometimes even later? Let us hear from two scholars who were not apparently even thinking of reincarnation in any form. The first will be from Lawrence Krauss in his book entitled *ATOM* in shortened summarized form:

I am drawn to the Ocean. . . . It still seems almost surreal to imagine that so much could have happened to make the simple act of my standing here possible. Can each of the atoms in the air I breathe really have gone through hell and back, braved the bitter cold of space, the brutal heat of stars, have crashed into the Earth, have dredged down below the continents and ocean floor merely to rise again? Have these atoms been a part of countless lives, and seen countless deaths? Will they travel throughout the cosmos? . . . I dive in, for a moment not knowing if I shall ever resurface. . . I know my atoms are likely to return from the ocean depths . . . and their future seems inevitably written, regardless of my own hopes and dreams. I am only a temporary abode, and my life is an inconsequential moment in their vast eternity.

I read Krauss's *ATOM* before completing *Our Quantum World and Reincarnation*. I saw no reason to use his material, interesting and beautifully written though it was, as I thought I had covered the point in a single paragraph.

I was somewhat at a loss however when one critic complained that I had included no explanation as to how the atoms from a deceased person, even entangled atoms, would enter the body of a fetus or new infant. I had indeed mentioned the almost eternal life expectancy of atoms projected by scientists, something like a billion times the projected life of our solar system; also the fact that entanglement would long outlast the life of any human if the atoms comprising the organs of memory were indeed in such an entangled state; how atoms from the deceased would, or could, become "part of the atmosphere, another human, a plant or animal, or travelers in space;" and that "atoms in our bodies were once in the sun, or other star, or a part of another animal of human" (p 117).

I had spelled out also that atoms in anyone's body may have previously been in the body of Alexander the Great or George Washington, or, on the other hand, from Attila the Hun or Nero.

But it was all in prose, in the midst of much other prose about many other issues. Krauss's work was also filled with prose, involving perhaps, even more issues. But in many places his work sings in poetry, such as the paragraphs summarized above from his final pages. For delivering a message with power, only music is more effective than poetry, but even Krauss's prose can sing and packs a potent punch well beyond the effective range of my bare mention of the few paragraphs from page 117 of my book.

His relevant passages however, most especially those of relevance to us, are indeed prose, facts, numbers, and formulas. We must remember our purpose here, namely to gauge the probability, or at least the possibility, that atoms, including some entangled with others, will find another temporary home in other bodies.

Some of the atoms will remain there during the new lifetimes; others will depart sooner. Judging by the findings of Dr. Ian Stevenson and his associates, whether the atoms in the organs involved with memories remain or not, the memories themselves will in 7 or 8 years often be replaced, if not entirely, then very largely, by the experiences of the new person. This is not reincarnation as most believers visualize it, but it is as much of it as can be confirmed scientifically by current research.

Remember also that Krauss is not concerned with reincarnation, and that neither the word nor concept of entanglement is touched upon by him. His interest is largely the imagined history of the life of a single atom. Fortunately for us, when he turns to recent history he continues not with the single atom, but with the vast conglomeration of them in the breath of air the reader takes as he reads. He happens to deal with atoms of oxygen, but pays due obeisance to the fact that all of the atoms of the many elements have histories more or less similar. We will later in this chapter see some similar observations about the atoms of carbon, another of the elements vital to life.

We continue with the reader and his breath of air. Krauss does the math, the details of which will interest few of us, and determines there is in each breath a half a liter in which there are 6×10^{21} oxygen atoms, a 6 followed by 20 zeros. He then finds 2.6 kilograms of new organic material in each square meter of forest each year, of which @80% is used to respire carbon dioxide and water back into the atmosphere; @20% is stored. He ultimately finds it reasonable to conclude that the oxygen *atoms* we now breathe are continuously redistributed throughout the atmosphere on a time frame of centuries. The result is that the *molecules* in every breath over the next several thousand years, becomes redistributed uniformly. So he concludes, we may be more connected to our past than we might have imagined.

As an aside, it might be noted that the sudden switch from 'atoms' to 'molecules,' the words italicized above, can be somewhat confusing. Free oxygen in most of the Earth's atmosphere consists of two atoms joined into a molecule. By its atomic structure a single oxygen atom is very active and will oxidize almost anything. In the lower atmosphere they freely join with each other. It is difficult to know, when Krauss speaks of numbers of atoms, if he speaks of two for each molecule, or, as in the above paragraph, the molecule seems to count as a singular item? Most likely the molecule counts as two atoms. We will assume that.

Krauss then turns to a specific and well known case, the moment Caesar was killed, and cried out "Et tu, Brute?" Krauss's following conclusion is based on the assumption that this involved 4 times as much oxygen as the $6x10^{21}$atoms. This, he says means that the makeup of oxygen from atoms in Caesar's last breath was 5 parts in 10^{22}, obviously an exceedingly small part. But if in every breath one takes today, one, breathes $6x10^{21}$oxygen atoms, the result is that we take on @ 3 of oxygen atoms from Caesar's last breath in each breath we take.

He also cites a fellow mathematician, who, following a different probabilistic route, concluded that there is less than 1% chance that none of the molecules we breathe came from Caesar's last breath. Phrased differently, there is at least a 99% chance one will.

This, says Krauss, is the good news. We have breathed, and continue to breathe the atoms of royalty. But without mercy Krauss takes us to what inevitably follows.

If the air in the atmosphere is shorter than a century, the same may be the destiny of the breath of Adolph Hitler. In this sense, says Krauss There may well be a "molecule" coming from every breath from every person who ever lived, unless the time has been too recent for the molecule recycling. *Molecule?* Well it is the molecule that is recycled, so we need not be confused by that.

But, if during respiration and photosynthesis, oxygen is linked from time to time with hydrogen to form water, there is the likelihood that at some point in its history every oxygen atom in our breaths was part of a water molecule (H_2O). This water molecule, he continues, has some probability of having been part of some excretion of someone who lived before. We may be breathing what was part of someone's urine or semen. He takes it further. The sweat of our parents' couplings, may be contained in the water we drink today. The same holds true, he says, for horse urine or pig feces, and we can carry this back to the beginning of life on Earth, when oxygen first built up in the atmosphere. There hence would be the chance that we breathe in some atom excreted by at least one of every species that ever lived.

Krauss then reminds us of something that will, very shortly, be of particular pertinence to us. It could be that you are not guaranteed to have a molecule from Caesar's dying breath in the air you breathe. It could instead be possible, or perhaps even plausible, that each oxygen atom in the breath you take has a unique history. "Some histories are exotic, and some or not."

In short, we are dealing with statistics, and we are dealing with chance. If statistics says we have the average chance of three molecules of air once breathed by Caesar, we should not be surprised if we learn that some have 15 and others have none.

More important: Is it not now self-evident how atoms of all types, including those that once comprised the organs of memory of another person might become part of a growing fetus? And we have seen the probability that such atoms may, and probably are, entangled with all atoms of such organs.

Quite apart from the recycling, we eat plants and foods that grow in the ground; food stuffs that have been nourished by the droppings of animals, or by the decaying of human waste. Not only oxygen, but nitrogen, calcium and all other elements we need and upon which we thrive, are furnished by the air we breathe, the fruits and vegetables and meat we eat, all made of atoms, distinguished from each other only by the numbers of protons and electrons they harbor. I would assume such knowledge is available to anyone who has any interest in the subject, and that the few lines I wrote about it, would have been sufficient. Apparently I was wrong

*   *   *

We will next hear from Primo Levi about the 'biography' of an atom of carbon in his *The Periodic Table*. According to him it is possible to claim that the arbitrary story he recites is true and that he could recite numerous other such biographies of other carbon atoms and that they would all be true. How so? There are, he claims, a significantly huge number of such atoms that one could always be found with a life history to match his invented one.

He says he could recount others that became *colors or perfumes in flowers; of others which, from tiny algae to small crustaceans to fish, and gradually return to the waters of the sea, in a perpetual, frightening round-dance of life and death, in which every devourer is immediately devoured.*

From there he turns to more civilized environments. Some of the atoms of carbon will settle in some archival document, or the canvas of some famous painter. Others will form a grain of pollen and have left their fossil imprint in the rocks for our curiosity. Still others descended to become part of the human seed, and "participated in the subtle process of division, duplication and fusion from which each of us is born."

The particular atom of carbon that Levi conjures lay for hundreds of millions of years as part of limestone, attached as it was to three atoms of oxygen and one of calcium. The only change it experienced was due to temperature variations. But then came man and his pickax. As the inventor of this narration Levi selects 1840 as the year in which a chunk of it was detached and sent to the lime kiln, "plunging it into the world of change to be roasted until separated from its calcium," but clinging to two of its three atoms of oxygen, thus becoming its first life as carbon dioxide. Its story was now tumultuous. Says Levi: It was caught by the wind, flung down on the earth, lifted ten kilometers high.

It travelled with the wind for eight years, on the sea and among the clouds, over forests, deserts, and ice. Then, says Levi, it stumbles into capture and the organic adventure.

The atom that Levi imagines, accompanied by its two atoms of oxygen, all in a gaseous state, he sees as being blown along a row of vines in the year 1848. It was brushed against a leaf, penetrated it and nailed there by a ray of the sun. It is all quite different from that "other organic chemistry," the cumbersome, ponderous work of man. This natural one was invented "two or three billion years ago by our silent sisters, the plants," Their scale is a millionth of a millimeter; their rhythm, a millionth of a second.

His atom of carbon enters a leaf and attaches to a large complicated molecule that activates it and also receives the "decisive message from the sky." The message is a packet of solar light that catches the leaf. It is separated from its oxygen and combines with hydrogen and with five companions enters the chain of life as a beautifully shaped structure and is dissolved in water. It is now part of a molecule of glucose. Our particular atom, per Levi, descends to become part of a grape, then wine, whereupon after sufficient exertion by the consumer it becomes glucose again, then lactic acid, and ultimately, breathed out as a new molecule of carbon dioxide, is returned to the atmosphere.

Levi's atom now travels far. It sails over the Apennines, and the Adriatic, Greece, the Aegean, Cyprus and Lebanon. It is now trapped for centuries in the trunk of a cedar. It could last there for five hundred years, but Levi chooses twenty. This puts us in the year 1868. A wood worm has dug a tunnel between the trunk and the bark. It forms a pupa, which is its inactive immature form between larva and adult stages. This one in the spring comes out as a gray moth.

The moth lays its eggs and dies. The cadaver lies in the undergrowth of the woods. Buried by dead leaves and the loam, it is safe from the weather; it has become a thing. But, says Levi, writing in 1984, the death of atoms, unlike ours, is never irrevocable. Were he writing today he might better have said that some atoms are destined to last perhaps a long as a billion billion years.

On the average, says Levi, every two hundred years every atom of carbon, not congealed by material such as limestone, coal, diamonds, or certain plastics, enters and reenters the cycle of life through photosynthesis. And beside the vegetable and animal worlds there are deposits of coal and petroleum. But, these too are the inheritance of photosynthesis carried out in distant epochs, making photosynthesis the sole path by which the sun's energy becomes chemically usable.

Levi has one more example for us. Perhaps it is not needed, but nonetheless highly relevant to our area of interest. It involves a glass of milk. Almost all of its links are acceptable to the human body. The one that concerns us here crosses the intestinal threshold and enters the bloodstream: it migrates, knocks at the door of a nerve cell, enters and supplants the carbon which was part of it. The cell belongs to a brain, his brain, says Levi, and the cell in question, and the atom in question, are in charge of my writing. "It is that which at this instant .. . makes my hand run along a certain path ... guides this hand of mine to impress on the paper this dot."

Is there any more graphic picture of how a cell, including an atom forming part of it, ones that with others contain the memories of a living human, might enter the body of a growing fetus, or in some rare cases, that of a grown adult? I have never, and do not now claim any of this to be proof that this is the way it happens. I can understand anyone's reluctance to accept it. I do not, however, understand any claim that there is no legitimate means by which it could happen, or requiring any more supportive suggestion than the writings of Krauss and Levi. At this point I am almost ready to ask any critic or skeptical reader: What don't you understand?

Krauss and Levi were both talking about single atoms. But entanglement considered, how many atoms would have to enter the organs of memory of any individual before a call of assembly goes out to the billions of others? I do not know, and neither, at this juncture, does anyone else. Even with single atoms, no one has ever taken a motion picture or video of its travels and flights, or been able to predict what atom will go where. As we think about such problems, let us not lose sight of the other mystery – reincarnation, or the inheritance of the memories of another, whichever anyone else may choose to call it.

Sophia Taylor, of Sheffield, is the author of a thoughtful article in the Guardian of September 27, 2011 entitled *where-were-my-atoms*. She has probably read Levi's *Periodic Table*. She concludes that no matter where they were, for certain "all the atoms that make up your body were forged billions of years ago in the fusion reactors at the core of now long-dead stars." Actually scientists say that hydrogen and helium atoms, the latter being the lightest of all, which comprise over 98% of matter in the universe, preceded star formation. All other atoms were created by the stars. Appropriately she quotes Carl Sagan's utterance "We are all stardust".

She also quotes from Hamlet, Act 4, scene iii: "A man may fish with the worm that hath eat of a king, and cat of the fish that hath fed of that worm." Meaning what?" asks King Claudius. Replies Hamlet, it is how "a King may go a progress through the guts of a beggar".

The cutting edge of science does, happily, bring out the poetic in those of us who have such talent, or genius. I envy them.

*   *   *

But there is one other matter, whose relevance is worth pondering. It is exemplified by the report published in a Science News Letter in August 1965. It reported on results of a test involving rats. One of the creatures was trained to go for food either at a light flash or at a sound signal. Chemicals from that rat's brain was injected into the brains of other rats, whereupon the infected rodents then "remembered" (the quote marks are from the Science news Letter) whether light or sound then meant food. Many scientists were said to believe that rats and humans have the same basic memory system and that humans might react in the same way with transplanted brain chemicals.

Previously an untrained flat worm ate a trained one, and a kind of "memory" (quote marks also from the paper) had been found, but the rat experiment was the first involving a higher form of life. Though flatworms are an extremely simple form of life compared to rats, they do have a rudimentary brain. Many scientists were said to believe that the mechanism for memory storage is the same for worms, rats, dogs and humans. They are said to believe further that the transfer of memory involves either DNA or RNA, or both.

Following this report, scientists from eight labs attempted to repeat the memory transplants. Their failure to do so was reported in *Science* magazine in August 1966 (vol153. P 658). Though no one claimed that the eight failures negated the validity of the one positive result that does indeed appear to be the clear implication.

We must acknowledge that elements of chance and pure statistics have been eliminated, at least in the opening steps, the injection of the chemicals from that rat's brain. But beyond the bare facts that the rats, and worms selected have been the recipients of chemicals from the brains of others of their species, how certain can we be that other factors beyond those may still be subject to nature's wheel of fortune?

Eric Kandel, who won a Nobel Prize in 2000 for his work on memory, has been quoted as saying that "Brain wiring is too intricate and complicated to be exactly replicated, and scientists are still learning about how memories are made, stored and retrieved." Does the fact of such intricacy and complication, rule out the possibility of chance that it could ever happen? Or does it, rather, rule out the possibility that it could happen more often than very seldom?

Ian Stevenson spent his mature life with a few colleagues scientifically showing, proving to more than a few, the efficacy of survival of memories. At most he can claim about 2500 cases, not all of which would he consider to be 'proved.' Would negative testing of another few thousand children disprove the validity of his work? It is doubtful that any but a very small portion of young children could ever be found to have valid memories of prior lives. The chances of such things may be very slim.

Over a thousand exoplanets have been discovered in the last few decades. None are believed to harbor life. There are very few who would suggest that this signifies that no other planets, except for ours, could do so. Much of nature is a waste. Millions of seeds by trees or plants, or human women never ripen or grow to maturity, or at all. Which few survive and which of the millions that remain do not, are largely matters of chance.

But what about the long accepted requirement for replication of the result of scientific experiment if such result is to be accepted? An article in *Science News Magazine* of December 26th 2015 by Tina Hesman Saey deals with problems arising from attempts at replication, and ways in which the problems might be minimized. Toward the end of the article, she acknowledges:

"Perfect reproductions might never be possible in biology and psychology, where variability among and between people, lab animals and cells, as well as unknown variables, influences the results."

The fact that one person sees a thing or an event particularly in fields such as biology cannot always be negated by the fact that a dozen other viewers making other attempts do not.

Chapter 10:

SINGULAR PROOFS

Cases of reincarnation, or memories of other lives, involve people, meaning an infinite variation of behaviors. One of the variations, exhibited on rare occasions, involves single recollections by the subjects, seemingly of little import, but which make much more convincing cases of possible reincarnation out of what had been very doubtful ones. We proceed here with two such cases, both from Italy. The first leaves perhaps some room for doubt, though to most minds, probably very little. The second will to most minds leave even less, not even by the high standards of Ian Stevenson.

Alessandrina Samonà, whom we will call Alessandrina II, the subject in this case, was the daughter of Carmelo Samonà, a, physician, practicing in Palermo, Italy. His wife, mother of both Alessandrina I and Alessandra II, was named Adele. Alessandrina I died of meningitis at about 5 years of age on March 15th 1910.

Three days later, the mother said, the deceased child appeared to her in a dream and told her that she had not left, but only withdrawn and would return as a baby. Three days later the mother had the same dream. Despite being told by a friend that this foretold of the reincarnation of the child, Adele, the mother, was unpersuaded and had never believed in reincarnation. Further, because of a miscarriage and complications over a year earlier, she did not believe she could become pregnant again.

Nevertheless on April 10th, 1910 Adele realized that she was pregnant. In August, an obstetrician examined Adele and told her she would have twins. They were born on November 22nd, 1910, possibly premature babies, which is less unusual in twins than in single births. One bore a remarkable resemblance to Alessandrina I and was named Alessandrina, whom we will call Alessandrina II as did Ian Stevenson.

The other twin was called Maria Pace. Their father Carmelo, the physician, noticed and described in writing, some physical similarities between the two Alessandrinas. They included hyperemia (the red eye) of the left eye, seborrhea (a mild dermatitis) of the right ear, and a slight asymmetry of the face. None of those attributes were found on Maria Pace.

By the time the twins were 2 ½ years old there were marked behavioral and temperamental differences between the twins, but pronounced similarities between Alessandrina II and the deceased Alessandrina I. Most of those similarities, truly remarkable as they were, could possibly, though not probably, be ascribed to similar genetics; Alessandrina II and Maria Pace were obviously not identical twins. But there is at least one episode that cannot be attributed to genetics or epigenetics. The account was part of a letter written to a colleague by Carmelo.

When the twins were 8 or 9 years old the parents spoke to the girls, Alessandrina II and Maria Pace, about taking them to Monreale, a town about 6 miles southwest of Palermo, the site of a magnificent Norman church. Adele, their mother, said "When you go to Monreale you will see some sights you have never seen before. Replied Alessandra II, "But, mama, I know Monreale. I have already seen it." Adele responded that she, Alessandra II, had never been to Monreale, to which the child replied:

"Yes I have. I have been there. Don't you remember that there is a great church there with a huge man on the roof with his arms spread apart?" She gestured with her arms as she spoke. Many churches could fit that description, but the child then continued, "And do you not remember how we went there with a lady who has horns, and we met a small priest who wore red.?"

Carmelo's letter continued, telling his correspondent that he and his wife had no memory of ever having spoken before about Monreale; that Maria Pace knew nothing about the place, and that they could make no sense of the lady with horns or the priest in red.

But a little later, wrote Carmelo, Adele suddenly remembered that a few months before Alessandra I died, they had made a trip to Monreale and taken the child with them. She also recalled that they had taken with them a friend who lived elsewhere and had come to Palermo to consult with doctors about some excrescences. These are distinct outgrowths on a human, or other living body, usually the result of some disease or abnormality. These were on the friend's forehead. Adele and Carmelo both also remembered that they met some orthodox priests who wore blue robes ornamented with red and that those details had much impressed their now deceased child.

Carmelo ended his narrative by pointing out that even if they "supposed" that one of them had spoken to Alessandrina II about the church it would not be believable that either he or his wife would have mentioned the "lady with horns" or "the priests dressed in red,' as those were features of no interest to them.

Stevenson ended his account by emphasizing that the suggestion of some critics that 'maternal impression' by the pregnant mother upon the daughter *in utero* was insufficient. It could not account for the detailed knowledge that Alessandrina II had of the visit by Alessandrina I to Monreale.

The entire doctrine of maternal impression, has long been discredited by the medical profession. Their skepticism was based largely on the wide belief that changes in traits can result only from mutations in the genes involved in procreation that occur before pregnancy. Changes in traits acquired during the life time of either parent, termed the 'inheritance of acquired characteristics,' was anathema, and at the time was rejected by the entire profession. Their typical example was the strong arms of a shoemaker or other father with muscles well developed through work, but does not result in an infant or young person with any unusual musculature..

However, to add somewhat to the previous mention of epigenetics, that branch of biological science was not advanced until the 1950s. It was adopted by a meeting of five specialists in the field in 2008, when it was defined as "a stable heritable trait that can result from changes in the chromosomes without alterations in the DNA sequence." The chromosomes are structures within the cells where DNA is located. This largely undercut, for many, the basis for the rejection of maternal impression.

Stevenson never questioned the possibility of maternal impression, but apparently never relied much on it. In various cases, he answered, and disagreed with critics who advanced the idea of that theory as an alternative to reincarnation.

Practically all of the other similarities in the two Alessandrinas, which we need not detail here, could have resulted from genetics or epigenetics, sufficient though they may have been to arouse suspicions of reincarnation. There are many other similar cases which Stevenson has rejected for lack of a 'smoking gun,' or something close to it, that we have here. Admittedly there is still room for an imaginative alternative for those determined to reject it. That is something for which the next case may have little, if any room at all.

*　*　*

So for a case that may be made somewhat more airtight as a result of one piece of evidence let us look at the case of **Laure Raynaud.** She was born in a French village near Amiens, France in 1868. The report of the case was first published by Dr. Gaston Durville in *Psychic Magazine* in January 1914, a year after Laure's death. Extracts were published in 1922 and 1924 by different authors, which Stevenson read. He had not seen the original report. He tells us that the two extracts agree in the essentials, though each contains a few details omitted by the other. In his narrative to us, he has drawn on both.

Laure's mother told Dr. Durville that as a young child Laure rejected the teachings of the Catholic Church, and insisted that after death one returns to earth in another body. As a result of the complications this caused, she stopped her overt challenges to the common beliefs of heaven and hell. At age 17 she became a healer and practiced first in Amiens, later in Paris where she studied magnetism, a precursor of hypnotism, in the school of Dr. Hector Durville. Stevenson is uncertain as to the relationship, if any, between him and Dr. Gaston Durville.

For the last two years of her life she worked for Dr. Gaston Durville in his clinic in Paris. She married in 1904, and died in 1913. Her husband told Dr. Gaston Durville that Laure had spoken to him about specific memories of a previous life from the time of their first acquaintance. By the time she began working for him she spoke freely about those memories to anyone who would listen. The doctor was a skeptic, but despite his skepticism, he did listen.

From the extracts of the original article, Dr. Stevenson states that she claimed the following details from the memories of a previous life: She lived in a sunny climate, most likely Italy; The house was much larger than ordinary houses; It had many large windows, the tops of which were arched; It was a two story house with another large terrace at the top; It was located in a large park with old trees; The ground sloped down in front of the house and upwards behind it; She had a chest disease and coughed much; Near the large house there were many small houses where workers lived. This life, she believed, had occurred about a century earlier, and she was about 25 years old when she died.

Laure's images of life there seemed very clear. She felt depressed and unstable. She remembered no names from that life, either of persons or places, but felt sure she could recognize the house should she ever see it again. In March 1913, the year of her death, that opportunity came. She was sent by Dr. Durville to help a patient in Genoa. Upon arriving in Turin she began to experience a familiarity with the countryside, a feeling that became more pronounced upon reaching that city.

She told her host in Genoa of her previous life and her wish to find the house she described. The host, Piaro Carlotti, took her to a house he thought matched her description on the outskirts of town. Laure said that this was not the house, but felt that the one she remembered was nearby. They drove further and came to a house that Laure said was "hers." It was owned by a prominent family. The house did correspond well with her description.

The dominant feature, according to Dr. Stevenson who later inspected a photograph of it taken in 1922, was the "unusually tall windows with arched upper ends. There was a large terrace around the lower level of the house and a small terrace at the top. Only two details did not seem to correspond. The photo showed that the house had three stories, though Stevenson explained that the sloping ground considered, one story may have been a "kind of a basement." Further the photograph does not illustrate the park and trees or show the slopes of the ground in front or behind it. One of the extracts of the original publication was in accordance with Laure's description. The other extract mentioned only the slope at the back, which was downward, contrary to Laure's description of it.

Thus far, despite the similarities, there was no way to prove that this was indeed the house Laure remembered. Who could tell how many features, tall windows, arched tops, terraces etc., disregarding even the minor discrepancies, could be found in other houses in the neighborhood? The lack of memory of names removed the usual means of verification. The similarities might well be coincidences, unlikely but not impossible.

But after Laure and Carlotti returned to Genoa Laure seemed to retrieve another memory, she had not had earlier:

She said that she was certain that in the previous life her body was not buried in the cemetery but in the church itself. Coming so late, it obviously was not a matter to which she seemed to attach much importance. But to Piaro Carlotti it was possibly of very much significance. Such church burials are not frequent; they are rare. He reported this recollection to Dr. Gaston Durville. The Dr., the confirmed skeptic who was not impressed with Laure's narrative or claims, initiated a search of the parish records. A correspondent in Genoa sent a record to Durville. It read:

October 23rd, 1809. Giovanna Spontini, widow of Benjamino Spontini, who lived for several years in her home, who was chronically ill and whose state of health recently gave rise to much concern after she caught a severe chill, died on the 21st of this month. She had been strengthened by all the offices of the church and today with our written permission and that of the mayor her body was brought with a private ceremony into the church of Notre-Dame-du Mont.

There comes a point at which coincidence is hard to accept as coincidence. There appears no opportunity whatever for Laure, born in France 1868, and where she thereafter lived and worked, to have learned about the obviously unusual case of burial in the church near Genoa, Italy in 1809. Nor did she seem to attach much importance to this memory when it surfaced. It required significant, and perhaps tedious, research to find such a case in the old records of the parish, and no one knew the name of the person for whom they were looking.

It is not uncommon for the subjects in cases of this nature to remember the events immediately following the death of the prior personality, even when they remember little else about 'the time in between.' We have seen one example in the case of the Druse boy in Chapter 4.

Though this and the Alessandrina case have in common a crucial piece of evidence, the cases are otherwise widely different, and it is pointless to look for similarities. The mother of the Alessandrinas knew exactly who the prior personality was, namely the predeceased daughter, and that they were dealing with one who had the memories of the other. Laure was Laure Raynaud, and there is no apparent relationship between her and Giovanna Spontini who died 59 years before Laure was born. Laure felt that she had lived in this home she remembered a century before her present life. Her memories of that life actually would have ended 59 years before, but could have begun much earlier.

As a result of the discovery of this one piece of evidence other matters seem to fit very well into the mold. One would be her insistence that reincarnation, her church notwithstanding, was a reality. Her depression may have been based on recollections of the early death of Giovanna, whose name she did not remember, and fear for own fate, something suggested by Stevenson himself.

Though Stevenson himself does not make much of the scarcity, almost uniqueness of an entombment in the church, the subject is worth a few observations.

Italy, location of the Vatican, has long been mostly Catholic. On May 2, 2012, there was promulgated the *Guidelines for Funerals and Burials in the Catholic Church*, a document that codified the centuries old accepted rules of Catholic burials. It provided that "Whenever possible, those who were part of the Catholic community are buried in a Catholic cemetery. As well as being a sacred place, it recalls the community of all the faithful, living and deceased." The document further states that "This service at the cemetery is the last farewell, in which the Christian community honors one of its members before the body is buried or entombed. With priest and mourners accompanying the body to the cemetery, the rite is celebrated at the grave or tomb or in a cemetery committal chapel." The cemetery committal chapel is, of course, not the church building, but a part of the cemetery.

Permission for entombment, or burial in the church is not frivolously granted. We can get a feel for the requirement by the findings in our case of the closeness to the church of the deceased, *Giovanna Spontini, widow of Benjamino Spontini*. Such an honor is obviously anything but routine. Inasmuch as only this one mention of church burial was sent to the doctor, it was apparently the only such case found.

For many reasons, historical and cultural, there are more such instances of church entombment in Florence than in other parts of Italy, and a look at the subject as practiced there may be instructive. Of all the tens of thousands that died there over the centuries, there are said to be only about 300 entombed in the most famous of its churches, the *di Santa Croce*. We obviously know little of most of these deceased persons but we can get a glimpse by looking at those among the most renowned among them:

They include Leon Battista Alberti, one of the town's best known architects, designer of many famous buildings; Dante Alighieri, the author of *The Divine Comedy*; Galileo Galilei, the great astronomer, first to turn a telescope on the moon; Michelangelo Buonarroti the renowned painter and sculptor; Gioacchino Rossini, the composer of thirty-seven operas, including the popular *Barber of Seville;* and Eugenio Barsanti, co-inventor of the internal combustion engine.

There of course are other churches. The Medici Chapels are entered through the crypt containing the tombs of who else but many of the Medici. The Chapel of Princes, heavily decorated with dark marbles, and semi-precious stones is the mausoleum of the Medici Grand Dukes. A New Sacristy contains two magnificent tombs sculpted by Michelangelo, the tomb of Giuliano, Duke of Nemours, beautified with the figures below his statue called Day and Night. The Tomb of Lorenzo, Duke of Urbino, has figures below called Evening and Dawn.

Finally, we should mention St. Ursula Convent: Beneath the flooring of their chapel, a burial was found that was tentatively identified as Lisa Gherardini Del Giocondo, the model for the Mona Lisa. This has still not been confirmed, but it is reportedly known that she was buried in that church.

What was the reaction of the skeptic Dr. Gaston Durville? "So, is this then a caused of reincarnation? I have to say that I know nothing about the subject, but I believe the reincarnation hypothesis is no more absurd than any other."

And Dr. Stevenson? After noting that the details described by Laure could probably be found in other Italian mansions of the Renaissance style, he added: "If however, we add to these details those of a chronically ill proprietress whose dead body was placed inside a church instead of a cemetery, the likelihood of coincidence diminishes substantially."

It does indeed.

Chapter 11:

STATISTICS AND THE 'TIME IN BETWEEN'

S tatistics can offer an overview of the entire field. Dr. Tucker mentions some significant ones from studies in the 21st century. One statistic will be of particular importance, but to keep it in context, let us see first an overview of the numbers he reported in his *Life Before Life*, pp 174 et. seq.

Depending on the poll, he claims, between 20% and 27% of both Americans and Europeans believe in reincarnation. A Harris poll in 2003 found that 21% of Christians in the U.S. believe in it. Most all of the believers no doubt think in terms of continuing their lives, with all mental, emotional and ethical characteristics in other bodies. In short it is the soul that survives. It is doubtful that they believe only in the limited form of it shown here by Stevenson and other researchers.

Mention has already been made, in Chapter 6 of the poll for Reuters of global citizens who believe in some form of afterlife, broken down to percentages by those who believe in an afterlife but not specifically in a heaven or hell, and those who believe we do go to heaven or hell. The findings are from a survey conducted in 23 countries among 18,829 adults. The numbers of totals used in figures cited in Chapter 6, namely 21% are not given.

The figures Tucker does give are interesting and informative in themselves. He cites for instance Stevenson's *Reincarnation and Biology: A Contribution to the Etiology of Birthmarks and Birth Defects*. It details 421 cases from India, 225 cases showing marks or defects with various pictures, and many of those were shown to have corresponding wounds on previous personalities. Of the 225 cases, 40, or 18%, have medical records confirming the match.

A surprising result arises from another statistic to the effect that although the number of statements from witnesses supported by their written records are almost 30% more frequent than statements of those relying only on memory, the percentage of statements proved to be correct was virtually the same in both groups, 76% of those with written records versus 78.4% relying on memory.

Tucker also cites Stevenson to the effect that when the subject and the previous personality were from the same village, the case is not usually as impressive as when subjects report memories of persons completely unknown to their family. No surprise there.

Of 973 cases from varying cultures, there were 195 cases where the same family was involved with both the previous and new personalities. In another 62 the two families had a close association before the case developed; in 115 a slight association. In 93 cases the subject's family knew of the previous personality but with no association. Of the 971 cases, 508 were 'stranger' cases with no contact or relationship between the two families. Of these, 239 were solved; 232 were unsolved, and a tentative identification was made in the remaining 37.

As previously mentioned, Stevenson explains his use of the term "solved" and "unsolved:" "When we satisfy ourselves that the child's statements correspond substantially correctly to the life of a particular person, we described the case as *solved*. Cases for which we fail to identify such a person we call unsolved."

Stevenson also compares certain results from studies of cases from Europe on the one hand, and the rest of the world on the other. He has often emphasized the comparative scarcity of European cases and the greater difficulties in investigating. Nonetheless the percentages are surprisingly close. The sampling involves 668 from "other countries," and 22 from Europe. The mean age of first speaking of the other life in other countries is 36.4 months; in Europe it is 35.9 months. The mean age of those ceasing to speak of it spontaneously is 85.6 months in other countries; 97.1 in Europe. The percentage of children mentioning the mode of death in other countries is 76.2, in Europe 84.6. The percentage of violent deaths of the previous personality in other countries is 68.8; for Europe, 73.7. There does seem to exist a universal pattern.

Satwant Pasricha has compiled figures dealing only with India from cases for whom information is available for the following particular feature. The name of the previous personality is recalled by 87.8%; by 12.2% it is not. Those who insist on being called by the previous personality's name total 11.5%; 88.5% do not insist. The mode of death is recalled by 66.7%; 33.3% do not recall it.

In a study, mostly from one of Stevenson's later publications, he found the average and median ages for the first declarations and behavior related to the previous personality to be 2.6 and 2.3 years respectively. The duration of major signs of 'personation,' was on average 6.9 years, with a median age of 7 years. Personation is a term used by Stevenson where outward signs include, among others, repeated expressions by the subject about the previous personality in the form of memories of events experienced or people known, emotional display appropriate to the memories, and habits and skills appropriate to the prior personality.

Of most interest perhaps is his findings on the ages of ending or diminution of memories. There were 13 subjects available for interview. They claimed either some fading, or complete amnesia. The earliest age mentioned was 14 years, the subject reporting persistence, but some fading was evidenced. The highest age mentioned was that of Marta Lorenz, whose case will be summarized shortly in these pages. Other than Marta the highest age was 27. She, at 54 years, could be classified as an outlier, claiming at that time, partial fading and partial retention of memories.

For all 13 the average age for total or partial loss, by my calculations was 23.8 years. Excluding Marta the average was 19.7.

* * *

Tucker also examined the results from a grouping of 1100 cases in which he compared the mode of death, where known, of the previous personality, namely those dying by natural means versus those by unnatural means. 'Unnatural' means, include drownings and violent deaths, such as murder, suicide, or accident. It was found that the two groups were similar in their propensity to talk about earthly events occurring after the death. The talk about 'another realm' however was more frequently spoken by those who died a natural death, 19%, than did those dying by unnatural means 13%.

The term death as used above, means 'actual death' and is differentiated from the previous personalities' funeral or handling of the remains. Of the 1100 subjects, only 69 made statements concerning those matters. We have two such cases reported in this essay, chapters 4 and 10.

There is a second dichotomy used by Dr. Tucker to compare the two groups: deaths that were unexpected at the time of death, even a natural death such as by heart attack, and those who expected death, even for a period as short as part of a day. Again there is no difference between the two groups about earthly events. But there is a difference with regard to statements about another realm. Cases in which the prior personality died suddenly are less likely to include statements about the other realm, 12%, than are those who did not die suddenly, 22%. In this latter group the longer the expectancy of death, the more likely and more numerous the statements about the other realm.

Dr. Tucker acknowledges that because the data base contains only 1100 cases the results are "preliminary in the extreme, but nonetheless statistically significant." We work with what we have.

Satwant Pasricha's figures for India only show that those who recall some details of other realms, presumably including earthly events as he does not distinguish between the two, constitute 18.5%; an equal percentage of 18.5% recall very few details, and 63% recall nothing from that period. Hence, the totals here for India alone are similar to the combined totals for the 1100 gathered worldwide.

As often happens, such numbers can raise more questions than they answer. One of these involves the question of why we hear so few cases of these children speaking of whatever else survives, between the death of the prior personality and birth of the subject, whether earthly events, or what we will call the 'other realm,' that is the abode during that period, wherever it might be. All of the subjects reported by investigators involve many aspects of the life of the prior personality. It is the sole means by which we can get a clue of a prior life. Yet proportionally few of them speak of the 'other realm' or of earthly events during the interval between the earlier death and the later 'rebirth.'

The relative paucity of such memories of the realm between death and 'rebirth,' versus the incorporation of many detailed memories from the prior life stand in sharp contrast with each other. The clearest memories from earthly events or the 'other realm' usually come from the time immediately after, or closely following the death of the prior personality, and seem quite similar to reports of those speaking of near-death experiences, something we have not touched on in this essay or in my book. We have seen at least two such cases of recollection of funeral or handling of remains following the death of the prior personality, both involving important facts that could be confirmed. One is that of the Druse boy identifying 'his' murderer, in chapter 4, and that of Laure Raynaud recalling 'her' burial in a church, rather than a cemetery, chapter 10. Both involve recollection of the 'earthly events,' not of the 'other realm.'

It appears however that there may be a relatively cogent explanation for the paucity of recollections from the time 'in between,' whether from another realm, or from an earthly one. It is one that has to do with the biologic function of formation, usually referred to as 'encoding,' and retrieval of memories on the one hand, as opposed to storage of memories on the other.

Our thought processes, including encoding of memories involves synapses, which are connections between neurons, which are cells of the brains. There are approximately 87 billion of them in each of us, the bodies of which are called soma. Each has a number of dendrites, long filaments with much complex branching, and a single axon which may be thousands of times the length of the soma. Each neuron may be connected to as many as 10,000 others, resulting to possibly 1000 trillion (or a quadrillion) synapses.

The encoding and retrieval of memories are both by electro-chemical means. The chemical are in the form of ions, atoms with an electrical charge, of sodium, potassium, chloride, and calcium within the cell. If the voltage across the cell's membrane changes significantly, an electrochemical nerve impulse is generated. The electrical activity can be measured and displayed as a brain wave.

The pulse travels along the axon and is transferred across a connection to a nearby neuron which receives it through its dendrites. This is the synapse. It is approximately 20 millionth of a millimeter. They sometimes occur through other combinations of axons and dendrites. A neuron fires approximately 5-50 times a second.

So perhaps before we consider memories from the time 'in between,' we should look at where and how long term memories are stored in the living brain. Until a few years ago, it seemed settled that the hippocampus was their repository. It was accepted, as stated in a paper by researchers at Sloan Kettering that those that have lost function or had major portions of the limbic system removed, but still have the hippocampus, have absolutely nothing but long-term memory and cannot record any new memories or functions.

However, a paper published in 2013, co-authored by a physician at the Max-Planck-Gesellschaft stated that recent research has suggested that the motor cortical circuits themselves, and not the hippocampus, is used as memory storage. The paper also stated that the 'where' and 'how' "memories are encoded in a nervous system is one of the most challenging questions in biological research . . . it is the cerebral cortex, not the hippocampus that is the storage site for some forms of memory."

Hence, there is still much uncertainty about the entire subject. However regardless of the locus of storage, it is relatively clear that encoding and retrieval are very complex matters, whereas storage, until retrieval is needed, is significantly less so. That fact seems itself to suggest the answer to the question of why so few memories of the time in between survive. The intricacies of thinking and encoding of memories, we might assume, are not present in atoms scattered about. They can communicate, but there is not the closeness, nor, probably the electrical discharge needed. The human brain, including formation, storage, and retrieval of memories is extraordinarily complex. It is possible that only when they become part of a functioning human body can the memories in long term storage be retrieved. But there can obviously be no retrieval of memories that were never formed, such as during the interval between death and 'rebirth'.

As part of another internet post the author wrote: "While we don't know where memories are stored. We do know that certain parts of the brain are responsible for recording, archiving and retrieving them. When special parts of the brain are injured, certain types of memories are unable to be accessed. But when these brain parts recover their function, the old memories are found intact. So where were they hiding?"

Perhaps another question left to be answered in this area is why even a small percentage of subjects recall the other realm or earthly events immediately following death. The question is germane even though the memories that do emanate from that period are sometimes curt, and often unsatisfactory. It would of course be guesswork, but not all biological processes are found to be 100% in everyone. Even the finest doctors are wrong in their projections on occasion, and some results of medical processes cannot yet be explained at all. We must leave it at that.

Chapter 12:

Two Predictions: Brazil and Alaska

Can persons, before death, even if not after, control to whom, or to which family they will be reborn? It is a difficult concept for most modern western people to accept. We have seen one such claim in the case of Alessandrina Samonà in Chapter 10. But there are more such cases, and we should be aware of that before deciding. We present now two cases, one from each of two areas, in both of which there is a relatively high proportion of believers in both reincarnation and other aspects of the 'paranormal.' Both were investigate d by Dr. Stevenson.

Marta Lorenz was born August 14th 1918 to Ida Lorenz, wife of F.V. Lorenz, the father of Marta. However our narrative begins about 1890 with the birth of one, Maria Januaria de Oliveiro, but always known by the name Sinhá, or Sinházina. She was the daughter of a prosperous rancher, C.J. Oliveiro, in southern Brazil.

Though Sinhá loved the life on her father's land she frequently visited the town of Dom Feliciano to see Ida Lorenz with whom she became very friendly. Two love affairs of Sinhá were thwarted by her father, one of the men committing suicide. Her resulting melancholy was such that her father arranged a trip of consolation for her to the city of Pelotas.

She there appeared to undertake a course of conduct amounting to her own suicide, including exposure to inclement weather and self-exhaustion. After an illness of a few months she died of tuberculosis. On her deathbed she told Ida Lorenz that she wanted to die and had tried to become infected. There was some discussion between the women in which Sinhá was said to have talked about a prediction of later being reborn as a daughter of Ida. Sinhá died at age 28 in October 1917.

On August 14th, 1918, ten months after the death of Sinhá, Ida gave birth to a daughter, which she and her husband named Marta. Marta, as a child of only one year, had shown affection for the father of Sinhá from the earlier life, as she considered it, in preference to a Mr. Valentin, when the two men visited the Lorenz house together. Valentin showed every sign of friendliness, but Marta, the one year old, went to the one she considered her previous father, C.J. de Oliveiro, stroked his beard and was heard by an older sister of Marta by six years, to say "hello, papa."

When Marta was 2 ½ years, she began to speak of events in the life of Sinhá. The first remark was made to her older sister, Lola, in a conversation later repeated by Lola to F.V. Lorenz, father of Marta. According to the account of Lorenz, Marta, at age 2 ½, was returning from the stream near the house with Lola. She asked her sister, "Carry me on your back."

Replied Lola, "You can walk well enough. I don't need to carry you," to which Marta answered "When I was big and you were small, I used to carry you often."

Very little conveys the innocence or carries the badge of spontaneous truth as do utterances such as this. The reversal of roles is perfectly natural to the child who sees nothing unusual in it. There are a number of similar cases from various part s of the world.

Lola laughed: "When were you big?" Replied Marta, "At the time I did not live here; I lived far from here where there were many cows, oxen, and oranges and where also there were animals like goats, but they were not goats." (They were actually sheep.) When the two sisters reached the house and Lola described the conversation to Lorenz, Lorenz addressed Marta: "My little daughter, I have never lived there where you say you have lived." The child replied, "Yes, but in those days I had other parents." There was then some other discussion about Negro employees, again evidencing spontaneity of matters that proved true and could not have been known to other members of this family. It is compelling evidence, but even more so is the listing of numerous other examples recorded by Lorenz.

Some of these statements included the following: Sinhá's father was older than Marta's father, Lorenz. He had a great beard and talked gruffly. Her father in the earlier life had a Negro female cook and a Negro servant boy he beat – once for not fetching water causing him to cry for help. Sinhá intervened and together they obtained water from a well. There was recognition of the name of Florinda de Almeida as a former sweetheart of hers. Sinhá had been godmother to F.V Lorenz's son, Carlos. When Ida Lorenz came to visit Sinhá, Sinhá would prepare coffee and wait in front of the house playing a phonograph she placed on a stone.

Sinhá had acquired her last illness on a trip to the city at carnival time where there were many masqueraders. On the return journey they were caught in a heavy rain and spent the night in an old house. An older foster sister of Marta, Ema Bolze Moreira had heard the accounts from both Sinhá and Marta. Despite the ten months lapse between the death of Sinhá and the birth of Marta, Emma said that both accounts were substantially the same.

The young Marta also recalled that Sinhá had given her godson Carlos two cows before she died. She also claimed that Sinhá had a white horse. It was actually called clay colored. It belonged to her father, but she rode it in preference to her own red horse. Sinhá and Ida once bought identical saddles on the same day.

There may justifiably appear here some confusion as to whether Marta considered her own identity to be that of herself as Marta, or whether she saw herself as a continuation of the deceased friend. Stevenson says flatly that "Marta identified herself with Sinhá completely." However she often began her sentences "When I was Sinhá." To further complicate matters, Stevenson also learned that she at other times began with "When I was big." This is not the only case that can, justifiably, confuse the reader as to the mindset of the subject.

Sinhá used to sit next to her father at meals. She had a white cat. She was buried in white with something on her head. Sinhá's father spoke harshly to the slaves. Slavery was abolished in Brazil in 1888, about two years before the birth of Sinhá, but as in in the U.S. abolition did not immediately change substantially the situation of blacks.

Stevenson's interviews with the family were in 1962. Marta had been suffering from attacks of bronchitis. He did not see Marta again until 1972. She was then 54 years old. He writes that she said she had forgotten much about her life as Sinhá, but had also remembered much. However it did not seem to him that her memories had undergone any additional fading in the prior ten years. Perhaps she had forgotten much, but Stevenson says she had obviously retained with "vivid clarity" many of the details of Sinhá's life.

He noted that she still suffered from attacks of bronchitis. She said that every time she caught a cold it went to her chest and larynx. The attacks occurred about four times a year and involved laryngitis. None of her siblings were so afflicted. Sinhá, had contacted tuberculosis from which she died, and Stevenson writes that before she died she could only speak in a faint whisper.

Stevenson expresses no opinion, but lets the subject fade away. He rarely trumpets any case as proof. In that I believe he sometimes fails his readers. This case strengthens one's belief in the possibility of reincarnation, or the existence of a third layer of inheritance. It would have been well to hear some judgment of his.

We have seen in Chapter 6 how Hungary had the highest percentage of believers in reincarnation as opposed to the more vague 'life after death.' A reminder that in second place, only 1 percentage point behind Hungary in that category, was Brazil.

*　　*　　*

Of all of the varied cultures from which Stevenson and others have found "reincarnation type" tales, hardly any would be more remote from European and Western culture in general than the Tlingit of southwest Alaska. It is pronounced as "klingit." Their primary home is the southern end of the Alaska coastline. This covers the narrow coastal strip of the continental shore along British Columbia. Today, Tlingit and Haida Central Council figures show a total Tlingit population of almost 17,000.

In addition to reincarnation they appear to accept, also without question, appearances of the dead in dreams, including announcing dreams, extra sensory perception, and other paranormal communications. They sound, it might ironically be said, like some of the most renowned of our scientists, particularly some quantum physicists. Needless to say their beliefs long antedate those of the scientists.

Their social structures and relationships are affected by the belief that all Tlingits are reincarnates of an ancestor. It is an interesting culture, and they are interesting people; they have suffered and endured.

Our case here is from this population, one researched and reported by Stevenson. The subject is **Corliss Chotkin Jr,** but we start with the death of **Victor Vincent**, the 'prior personality,' a full blooded Tlingit. He died in the spring of 1946 in the town of Angoon.

During the last years of his life Victor had become quite close to a niece, Mrs. Corliss Chotkin, Sr., the daughter of his sister. He had often stayed with the niece and her husband in Sitka, and always felt welcome. On one such visit a year before his death he said to his niece that he was coming back as her next son and that he hoped he wouldn't be stuttering as much then. He also said that "your son" will have these scars, pulled up his shirt and showed her a scar on his back.

It was a clearly seen scar from an operation some years previously, the small round holes from the stitching still showing. Vincent also pointed to a scar on the right side of the base of his nose. This scar had also followed an operation. Vincent assured his niece, that he knew he would have a good home as she wouldn't be going off and getting drunk, alluding to several alcoholics in the family.

About 18 months after the death of Victor Vincent, Mrs. Corliss Chotkin, Sr., his niece, gave birth, on December 15th 1947 to a boy, who was named after his father, Corliss Chotkin, Jr. We can overlook without comment that a Sr. married a Jr. and that their son is also Corliss Chotkin, Jr.

It was soon noticed that the newborn had two marks on his body, shaped and located precisely the same as the scars pointed out by Victor Vincent. But the mark at the nose's root had moved down by 1962, when the boy was 15, and lay on the right side of the nose. The discoloration was reduced and the mark on the back was more like an operative scar and was markedly colored, all matters noted by Dr. Stevenson.

Along the margins of that scar were several small round marks, four of which lined up like a stitch wound. Those on the other side were not so definite. The mark itself had moved downward since the boy's birth and had become more obviously colored. Mrs. Chotkin attributed the color change to frequent scratching by Corliss, which led to inflammation and distortion of the shape.

When the boy was able to talk, family members tried to teach him to say his name. When at 13 months the mother tried to so instruct him, he impatiently replied, "Don't you know me? I'm Kahkody." It was the tribal name of Victor Vincent and was spoken with "an excellent Tlingit accent." As a result of Corliss's utterance, he was given the tribal name of his uncle.

As Corliss grew older there were a string of recognitions. At age 2, he spontaneously recognized in Sitka a stepdaughter of Victor Vincent and correctly greeted her as Suzi. The meeting was unexpected by all parties and no one had mentioned her name before Corliss spotted and greeted her. Also at age 2 Corliss recognized William, Vincent's son. Corliss spotted him also on the streets of Sitka and said "There is William, my son." At age 3 he correctly recognized Vincent's widow out of a crowd and said "That's the old lady," and "There's Rose."

In another occasion, as he was playing in the streets he recognized a friend of Vincent, Mrs. Alice Roberts, and called to her by her usual pet name. His mother was not with him, he was quite alone, as he was upon seeing one other friend of Vincent. She was present however for two other occasions. One was when he saw friends of Vincent from the town of Angoon. Mrs. Chotkin said that Corliss had recognized still other persons known to Vincent, but she could recall no details. All of these recognitions occurred before Corliss was 6.

Corliss also narrated two episodes in Vincent's life that, she feels, he could not have learned from normal sources. One involved the breakdown of a boat engine, for which he attracted the Coast Guard by changing uniforms and rowing. This brought attention to him and he had personnel from the Coast Guard contact someone for help. The second story involves a visit by Mrs. Chotkin and Corliss to the home where Mrs. Chotkin and Vincent once lived. Corliss pointed out a room and said "When the old lady and I used to visit you, we slept in the bedroom there.

This seemed to Mrs. Chotkin quite extraordinary, all the more so as the building had long been renovated for other purposes, and no rooms could easily be recognized as bedrooms. But the room Corliss had identified was in fact occupied by Vincent and his wife when they visited the Chotkins.

Beginning at age 9 Corliss made progressively fewer statements about a prior life. At the time of his interview with Stevenson in 1962, at the age of 15 he claimed to remember nothing.

Among the matters that impressed Mrs. Chotkin were the way Corliss insisted on combing his hair, a particular style, identical to Vincent's despite her attempts to change it. Corliss's severe stutter when young, persisted until age 10, and had apparently disappeared by the 1962 meeting with Stevenson. He was, like Vincent, a devoutly religious person, and both were skilled with boats and engines. Vincent had been left handed, as was Corliss until his school teacher persisted in teaching him to use the right.

Though Stevenson never precisely so wrote, his efforts to prove or disprove the correspondence of birthmarks on Corliss to marks on Vincent were not satisfactorily completed. He seems convinced however that the marks on Corliss were congenital, and not post-natal. Many of the witnesses who could have helped corroborate much of the evidence from Corliss and his mother were unavailable either through death or failing memories accompanying advancing age. Even after his later trip to Alaska, Stevenson was convinced of the honesty and generally in the accuracy of the mother, Mrs. Chotkin.

Stevenson, as was his custom, met again with Corliss and his family in 1965 and in 1972, when Corliss was 25. The meetings revealed Corliss to be a man of average abilities and qualities, failure in high school, military service in Viet Nam, and injury, not in combat, that left him partially deaf. The birthmark on his back had been removed surgically due to its persistent itching. In 1972 only the scar of the operation remained and it had healed well.

It will be recalled that Victor Vincent had wanted to be reborn without the stutter with which he was severely afflicted. In the 1972 meeting it appeared that to some extent it still lingered particularly when he was emotionally disturbed. His mother said he stuttered much less than Victor Vincent who "stuttered all the time."

Victor Vincent had been devoutly religious, active in missionary work and a major in the Salvation Army. Corliss was interested in religion in childhood and adolescence, but attributed the widespread use of drugs and miseries of war to failures or weaknesses of religion. Upon his return to Sitka and civilian life he had a distressing experience with a religious group, the details of which Stevenson does not tell us, that turned him completely away from formal religion.

Stevenson effectively answers arguments, or potential arguments of others, who account for the rather impressive evidence, including other types of paranormal processes such as extra sensory perception. Most readers, except those unwilling to accept reincarnation at all, will undoubtedly agree with Stevenson that the case is exactly what it appears to be: a case of reincarnation, or the limited type here considered. The possibility of announcing dreams and a choice of the prior personality cannot always be accepted even by believers in reincarnation, especially the limited form of it proven by Stevenson. Though questionable to many, this feature is not unique to this case, nor to the others such as the case of Alessandrina Samonà in chapter 10.

Chapter 13:

Black Holes, Worm Holes, Holograms, and the Arrow of Time

So physicists know that particles can get entangled and that if they do, certain complementary characteristics change in lockstep. The physicists acknowledge they do not know *how* that happens. Less often mentioned, has been an even more tantalizing matter, namely *why* does that happen?

In the overall scheme of things, what is the necessity for it? Certainly it cannot exist just to have some infants born with memories of other lives, memories that ordinarily last no more than 7 or 8 years, and which are often gone by the early teens. They last, that is, until the subject realizes that the memories are of little significance, and that what is important is the here and now. This is not meant to minimize the importance of such a scientific breakthrough, or the work of Dr. Ian Stevenson and others. It is indeed highly important, more for what it may portend than for what has thus far been shown, whatever may be the scientific explanation of their work, whether entanglement or something else.

Could entanglement be here merely to allow our scientists and technicians to use this feature of particles to make new gadgets, or to challenge our inventiveness, or to improve our technology? Could fate really be playing such a game with us?

Happily, in recent decades a number of physicists have been speculating and developing the beginnings of proof of some possible impact of entanglement, not on our technology or inventiveness, but on the nature of the universe. The subject of course has little relevance to our main topic, reincarnation, or the survival of memories even for a short period, or near death experiences, and other matters heretofore beyond the reach of science. But the subject of the basic function of entanglement is nonetheless irresistible. At the very least it may serve to keep things in perspective.

It does so because the speculation and beginnings of mathematical proof of the conjectured purpose has led to one thing that is rather shocking in its scope, and another that is highly significant. Let us start with the shocking one.

According to Leonard Susskind of Stanford University, entanglement was creating a "spatial connectivity that sewed space together." John Preskill, a theoretical physicist at Caltech summed up the theory as it was presented in an article in *Science News* of October 17, 2015: "Everything points in a really compelling way to space being emergent from deep underlying physics that has to do with entanglement."

Cited in the *Science News* article was also a paper published in 2009 by Mark Van Raamsdonk, also a theoretical physicist. He offered the view that "space without entanglement could not hold itself together," and that "quantum entanglement is the needle that stitches together the tapestry of cosmic spacetime."

Spacetime is a term, used in describing Einstein's view of the universe as three dimensions of space and one of time. To simplify, perhaps over simplify, in his intuitive breakthrough Einstein postulated that space was something as opposed to nothingness; that instead of a great void, it was a substance that could bend, and according to the relative view of speed by the relatively stationary observer, could shrink or elongate. Time also varied with relative speed and as time dilated with speed, length shortened in the view of that stationary observer. In his general relativity theory he also explained gravity as being caused by a bending of space around objects.

The *Science News* article further cited a 2012 paper presenting a paradox about entangled particles, one inside a black hole, the other outside. This was the impetus for the entry of the idea of 'wormholes,' tunnels in spacetime, earlier imagined by Einstein and physicist Nathan Rosen in 1935.

A wormhole was imagined, but thus far never seen. Many other of Einstein's ideas were conjectured by him but not verified by observation or experiment for years or decades. The wormhole as envisioned is essentially a shortcut connecting two separate points in space, possibly anywhere from a few feet to billions of lightyears, a tunnel with its two ends, each at separate points in space.

We also encounter here the possible role of holograms. The idea of Van Raamsdonk was rooted in the mathematics of the holographic principle to the effect that the boundary enclosing a volume of space can contain all of the information about what is inside of that space.

What is in three dimensions, in short, can be represented by two dimensions, something we all know, but was now applied to black holes. That insight was originally conceived in 2009.

The primary proponent of the idea that the holographic principle might tell us much about the universe, was Juan Maldacena, a theoretical physicist at Harvard. His idea had jelled in 1997, 12 years before Van Raamsdonk's paper. Maldacena was performing experiments designed to show whether the outside of two cans, each representing black holes, could be connected by a wormhole, the theoretical tunnel. Looking for a way to create the equivalent of the cans' labels, representing the boundary of black holes, he realized that the solution might lie with entanglement. Without entanglement he wrote, all of space would 'atomize.' This insight came in 2001, 8 years before Raamsdonk's paper.

He demonstrated that by entangling particles on each of two cans' labels with each other, he could describe the wormhole connection perfectly in the language of quantum physics. In the context of the holographic principle, entanglement was tying chunks of spacetime together.

It was Maldacena's work that impelled Van Raamsdonk to investigate what role entanglement might actually play in shaping spacetime. He tried to imagine the purist, emptiest disk of space possible. He soon realized however that quantum mechanics dictated that even so-called 'empty space' was filled with pairs of tiny, and extremely short-lived pairs of 'virtual' particles and that those particle pairs were entangled. Some of the pairs are anti-particles, and consequently they and the pairs of positively charged particles annihilate each other in a small fraction of a second.

It might be noted parenthetically that these particles that are believed to permeate space may ultimately prove of great significance themselves. Not least of the benefits, according to some scientists, may be a tremendous source of energy, rending current fuels outmoded and unnecessary. It must also be noted that these ephemeral particles are one more thing that no one has yet seen.

Returning to the imaginary disk of space, Van Raamsdonk mathematically bisected the imagined holographic label by an equally imaginary line. He then demonstrated that by severing, again mathematically, the entanglement between particles on one half of the holographic label and those on the other half, the surfaces of the hypothetic space disk started to split in half. "It was," he said "as if the entangled particles were hooks that kept the space and time in place; without them, spacetime pulled itself apart."

As he decreased the degree of entanglement, the portion of the space got thinner, and he noted and suggested that the "origin of having space at all is having this entanglement."

Maldacena, now at Princeton, wondered about another controversy raised in the 1990s. It began twenty years previously when Stephen Hawking had shown that entangled particles can get split up at a black hole's 'event horizon,' about which we will hear more a little later. One particle falls into the black hole, the other escapes. Hawking mathematically found that eventually the black hole would evaporate out of existence. Those escaping particles are known as Hawking radiation.

But in our particular context, this situation raised a conundrum. It is a law of quantum physics that complete destruction of information is forbidden. It may be scrambled up, such as in the burning of a book, but the atoms survive and theoretically (very) the information can again be pieced together from the smoke and ashes – just as, we might say, memories can perhaps be recreated when the atoms survive destruction of the human to whom they belonged.

It should be noted at this point that the word 'information' can have many different meanings. As used in the context of this subject matter it refers to the characteristics of particles, or to entangled particles jointly.

Returning to our present matter of interest, there arose a controversy as to whether the information in a black hole was just scrambled, or was it lost. It was Susskind who claimed that the information that fell into the hole went no further than the two-dimensional 'event horizon.'

That event horizon is the point below which nothing, not even light, can escape, the reason for which it is called the *black* hole. The gravity associated with the hole is extremely strong. Every cosmic body has an escape velocity, the speed needed to escape that body's gravity. For planet Earth it is @ 25,000 mph. Light, even at 186,000 miles per second cannot escape a black hole.

The black hole can be the remains of a collapsed star if the star before collapse was at least almost two or three times the size of our sun. Theoretically however any object could be a black hole if reduced in volume, and thus increased in density to the extent that its escape velocity would equal or exceed the speed of light.

Time is also warped by such gravity. By the laws of special relativity, the greater the speed, the greater the dilation of time as seen by relatively stationary objects such as ourselves. Speed of an object reaching the event horizon would seem to outside observers on Earth to stop, or to slow to an unimaginable degree, requiring perhaps billions of years to enter. A fictional rider on the object would see the Earth as spinning rapidly, around its axis and around the sun, and centuries, by his observation, happening in seconds.

Turning once again to our point of interest, it was Susskind who said that the event horizon encoded everything inside like a hologram. How so? From the view of an outsider, such as humans on Earth, the object and its information seemingly teeter on the edge, and hence the horizon becomes to us a holographic boundary, containing all the information about the space inside the black hole.

To interject one more explanation about holograms: For many years, other scientists have compared the universe to a hologram for a different reason and in a different way. The film that encapsulates the picture that appears as a hologram will appear as three-dimensional. It can be cut into pieces, even small pieces and each will show, not a small part of the original, but the entire original, only smaller. Those who believe in the underlying unity of all the universe, animists, for instance, are most apt to be counted among those seeing the universe as a giant hologram. Though this is irrelevant to our subject, it might save some confusion to those doing further reading on the hologram subject.

Back to our main path, there was still one more big surprise in store. It was Einstein who had proposed both entanglement and wormholes, as early as the 1930s, though he considered entanglement to be merely a thought exercise. It was intended to show just how short the quantum theory was of reality, never dreaming that this outrageous result could be a reality. But with the most recent advances in mathematical thinking, both worm holes and entanglement seem to be the same thing, or at least serve the same purpose.

In the 1990s a group of scientist in Santa Barbara suggested that black holes might not, in effect, have insides at all. The reason? A 'firewall just inside the horizon would fry anything attempting to enter. Why a firewall? Because relativity predicts that particles coming out of a black hole must be entangled with those falling into it. If they were not, if there were a break in the entanglement at the horizon, the result would be a discontinuity in space, a pileup of energy, or in short, a firewall.

Yet for information not be lost those particles coming out must be entangled with those that left long ago as 'Hawkins' radiation. But it could not be both. They could not be still entangled with the particles inside and at the same time with the Hawking's radiation outside. Each would have to be 'maximally' entangled for such a result as a firewall to occur, and maximally entangled particles can be entangled only with one other particle. It is 'monogamous.'

It was later explained by the same authors that it is only *maximally* entangled particles that could have no more than one entanglement and must remain monogamous. Hence that proviso would have no effect on the many multiple entanglements we have heard about and others that we have conjectured. An article in *Science News,* for January 9th, 2016, for a recent example, described a means of quantification of "a mysterious bond when it is shared by several particles rather than just two."

The suggested solution to the problem presented here were the wormholes. To them the firewall would be no impediment. Particles on each side would be connected by a wormhole that would link the interior of the black hole directly to the partner particles that left long ago as 'Hawking' radiation. There would be no discontinuity in space, no pileup of energy, and hence no firewall.

The notion of entanglement holding space together is seemingly gaining ground among the physicists and mathematicians. However there is still much controversy over firewalls, wormholes and certain characteristics of event horizons, let alone salvation versus destruction of information. "At first whiff," Dr. Preskill wrote in a recent chapter post about worm holes, the Maldacena-Susskind conjecture "may smell fresh and sweet, but it will have to ripen on the shelf for a while." He added, "For now, wormhole lovers can relish the possibilities."

Even Dr. Maldacena and Dr. Susskind both admit that legitimate questions are being raised. Few of their colleagues are convinced yet that it has been formulated in sufficient detail, let alone that it can solve the firewall paradox. "All I can say," Dr. Susskind said in an e-mail on the eve of a firewall workshop "is that no one has a completely solid case and that certainly includes me. Time will tell.""

According to the late Dr. Edward Polchinski, a UC Santa Barbara physical theorist, "My current thinking is that all the arguments that we are having are the kind of arguments that you make when you don't have a theory." We need a more complete theory of gravity, he concluded. "Maybe 'space-time from entanglement' is the right place to start," he wrote. "I am not sure."

Dr. Raphael Bousso, also a UC theoretical physicist, at Berkeley, in an e-mail to Dr. Maldacena, was skeptical that the wormholes would eliminate firewalls. "My own view is that it's time to move on, accept, and actually understand firewalls," he said. After all, he added, there's no principle of nonviolence in the universe, except for Einstein's equivalence principle, which says the black hole's horizon is not a special place. But maybe it is, after all.

The firewall paradox," he said, "tells us that the conceptual cost of getting information back out of a black hole is even more revolutionary than most of us had believed." Theoretical physics at this level is still very much a work in progress. But in the magnitude of its scope and importance, it may leave even reincarnation trailing in the dust.

*　　*　　*

We turn now to the second, somewhat less portentous suggestion about the function of entanglement.

It was in 1927 that Sir Arthur Eddington described the gradual dispersal of energy as an irreversible 'arrow of time.' It seems like an apt description. A hot cup of coffee cools to the approximate temperature of the room; the room does not heat up to the temperature of the coffee. Neglected structures crumble in time; the parts do not recombine into a new building. A dropped dish or jar breaks into pieces. It does not put itself together again without the use of more energy. The entire universe is sinking into a state of uniform lethargy known as thermal equilibrium, meaning temperatures will ultimately even out.

Perhaps it is all part of the laws of physics. Or is it? On the contrary it seems that the physicists are bewildered by the fact that the underlying laws of classical physics do not seem to support the one way direction of time's arrow, something we all see and accept, and which most of us would never seek to question, or even to think about. But the same laws would permit the reverse to occur as well. So why is time's arrow such a one-way thing? Since the advent of the science of thermodynamics in the middle of the 19th Century, the spread of energy to an equilibrium was attributed to the formulation of statistics involving the trajectories of particles which showed that over time equilibrium would be approached.

In recent decades, there has been new thinking to explain the result. The fundamental source for the arrow of time, say some physicists, is the fact that objects change toward equilibrium because of the way particles interact. They become intertwined by the 'strange process' as they used to describe it, called 'entanglement,' the same process that we have viewed from another perspective, namely, a limited form of reincarnation.

A paper in *Physical Review* published in 2009 proposed that entanglement caused evolution toward equilibrium, namely a state of uniform energy distribution. The authors claimed it would happen in an infinite amount of time. In 2012 the theory was refined to show that equilibrium would occur through entanglement in a finite period of time. Still later it was calculated that most physical systems reach equilibrium rapidly, in proportion to their sizes.

It all began with the idea that at the quantum level things are inherently uncertain. In the 'pure state' they can only be described by probabilities such as spinning, clockwise or counter clockwise. There is no 'true' state of the particle, says the Irish physicist, John Bell. The probabilities are the only reality. When two particles become entangled, a more complicated probability distribution describes both particles together.

The entanglement, for instance, might dictate that the particles spin in opposite directions. In that case the two could be any distance apart and the spin of each would remain correlated with the other. It is now the system that is in the pure state. The state of each particle is still unknown, though knowing one will also describe the other. If we find that one entangled electron spins right for instance the other will spin left.

It was first claimed by physicist Seth Lloyd in the 1980s that with the increasingly large numbers of entangled particles the spreading quantum uncertainty could replace the old classical proofs as the true explanation for the arrow of time. The particles thus can be said to have gradually lost their identity and become as pawns of the collective group. Eventually, says Lloyd, the correlations will contain all the information, the individual particles, none. At that point, he claimed, the particles will have arrived at a state of equilibrium and their states stopped changing, as did the coffee that had cooled to room temperature. "The arrow of time is an arrow of increasing correlations."

There is ongoing research and ongoing thinking and speculation. Much is not yet answered. As stated in the *Physical Review* article, "The new approach has yet to make headway as a tool for parsing the thermodynamic properties of specific things, like coffee, glass, or exotic states of matter." Also, we might specify, decay of buildings or broken dishes. The article also mentioned that some traditional thermodynamicists claimed to be only vaguely aware of this new approach.

There must indeed be much more to investigate, but it seems to be at least worthy of the attention it is getting to a growing degree.

Chapter 14:

WWII, Two English Boys and a Finn

Not all cases of reincarnation are pleasant experiences. Some memories from other lives are decidedly most unpleasant. We have already seen a little of it. In this chapter we will see much more. We end our summaries of such cases with three boys, all born more than 15 years after the end of World War II and the Holocaust. All three involve detailed descriptions of events, places, and things that they could not have learned through normal channels. One had the memories of a German fighter pilot. The other two were born with memories of children murdered in Nazi concentration camps.

Daniel Lewis was born September 1st, 1970 in Chester, England. His mother, Sylvia Lewis, was born in Wales. The father was probably not her husband, Jason, but more likely one, Jacob Rose, a Jewish man with whom she had had a two year affair. Sylvia and Jason later divorced. Jacob soon cut himself off completely from any contact with Daniel or Sylvia. The names of the principals in this case are changed in order to preserve their privacy.

Stevenson first met Daniel in 1998, a young man of 28. He told Stevenson that he remembered as a child meeting Jacob, and that he had a strong liking for him something he never felt for his legal father, Jason.

When Daniel was only a few years old, he began to waken during the night and exhibit symptoms of terror, including trembling. He began his writing going from right to left, contrary to what he was being taught, and showed a familiarity with Jewish customs that his mother knew he could not have learned from any normal sources. He was 11 before he learned to write left to right. While still quite young, he asked his mother whether food she was serving had blood in it.

At age 9, Daniel, with Sylvia and her second husband, visited another city where Daniel saw a building that looked like a church, but Daniel remarked: "They wear caps there." His step-father then said it was a synagogue. According to Sylvia, there were no people going in or out of it at the time.

More unsettling were scenes he described like memories of concentration camps and the slaughter of the Jews during the Holocaust.

In his nightmares he saw large dark and deep holes in which he could see bodies. He feared falling into one. He saw people with guns and he could smell the stench of dead bodies. He said the people there were like skeletons. They were bald and had no food. They were sitting around doing nothing. They wore "stripey things." Daniel often said "I'm worried for the other people. Why did it have to happen? Why did it have to happen?"

When visiting with Sylvia an aunt who cooked with gas, he became aware of the odor, and complained that "it going to smother me."

He also spoke, while awake, of "Walking around. Prisoner of war things." He said that people lived in wooden huts and that the people who lived there were prisoners whom he thought were Jewish people. When he drew, he always included stars though at times he seemed afraid of them. When he was about 12, shopping with his mother, he began to cry and ran out of the shop. He explained that he was afraid of a necklace he saw. It had a star sign and "It was beckoning to me." He continued to talk about the necklace and the star for some time. He also had a "hatred" for the color yellow.

He was fearful of camps. He was six when Sylvia suggested they take a vacation at one. Daniel adamantly refused. Sylvia patiently explained that people could have pleasant times at camps. Daniel was vehement: "No. there is no happiness there. People are caged in and cold, hungry and frightened. They'll never get out."

These symptoms continued throughout his childhood. In the summer of 1982, Sylvia learned of the work of Ian Stevenson and in September wrote him a long letter detailing the relevant history of Daniel. Stevenson had wanted to meet Jacob, Sylvia's first husband, but the attempt was unsuccessful. He had wanted to ask Jacob if any relative of his had been a victim of the Holocaust.

Ten years later, Stevenson, again contacted Sylvia hoping that both she and Daniel would agree to meet with him. Sylvia did agree and the three way meeting took place on October 16th 1998. Stevenson had a long meeting with Sylvia, a short one with Daniel, then 28. Unlike most other subjects whose memories began in early childhood, lasting only about 5 or 6 years, Daniel's persisted into adulthood. He, at this time, spoke of his memories as his "previous life," whereas he earlier spoke of them in the third person.

Also during that meeting Stevenson learned that Daniel was studying to become a nurse, and seemed "serious, but not troubled." From later correspondence with Sylvia he was informed that nightmares still troubled Daniel and that he felt he might "get it out of his system" if he went to Auschwitz, a name he had not used in childhood, but had learned through normal sources.

Stevenson said of this case that he believed it could not be explained by environmental influences or genetic inheritance. "Even the most ardent geneticist would not suggest that genes would transmit the habit of reading and writing from right to left, concern about whether the food had blood in it, and images of a concentration camp."

The ardent geneticists would most probably be the least likely to suggest that genetics was involved in such traits. Those are things most likely learned from the environment. But in Daniel's case, there was no such environment from which to learn them.

* * *

Stevenson has emphasized certain difficulties investigating cases from Western Europe and the lower 48 states of the U.S. The percentages of the populations expressing belief in reincarnation is substantially lower in those areas than elsewhere. Among other problems that bedevil most all cases from those areas, namely lack of willingness of possible witnesses to cooperate.

Stevenson terms the following case of **Christopher Ellis** as 'unsolved,' but apparently he felt it of sufficient interest to include it in one of his many volumes of cases. "Unsolved," as explained elsewhere means inability to prove the identity of the 'prior personality.' Whatever the reader may make of this short summary of it, it is deemed worth reading, as readers can make their own conclusions, and set individual standards of proof. Stevenson, despite his ultra-cautious approach, was obviously impressed with it. The names in this case have been changed to protect the privacy of the principals.

Christopher Ellis was born on December 29th 1972, in Middlesbrough, England. His parents were Jack and Veronica Ellis. Jack was a bus driver. Christopher had a sister 5 years older and a brother 11 years older. Christopher's very early words, when he could first make himself understood at the age of two, conveyed "I crashed a plane through a window" a thought he repeated often, gradually adding details. Also, according to his parents, between 2 and 3 years of age he drew sketches of airplanes with swastikas, though the parents noted that they were reversed. He also later drew an eagle, which his parents likened to a "German Eagle."

One of the details that Veronica recalled was that his plane had been a Messerschmitt and a bomber, adding the type as 101, or 104. But as of the interview with Stevenson in 1983, she could not remember which. It should be noted that Messerschmitts were best known as fighters, but late in the war there was a Messerschmitt bomber with the number 262, an early jet used as both fighter and low level bomber.

He said that he was on a bombing mission over England. He added other details, and early on he spontaneously gave the Nazi salute and mimicked the goose-step walk of German soldiers. When standing, he always stood erect, hands at his side. Though the initial drawings he made were crude, to his parents they clearly represented what they saw in them. As he became older, he drew with increasing skill. At the age of six he made sketches of the panel of a pilot's cockpit, and described the functions of the gauges He also showed an affinity for Germany and various behaviors that they saw as German.

Beginning in1982, when Christopher was ten, there was some publicity in an English magazine and a German newspaper concerning him, his behavior, and talk of an earlier life. Christopher's schoolmates treated him, not surprisingly, with cruel teasing and mockery, imitating his goose step type walking and calling him a Nazi. No doubt this was a contributing factor in his cessation of such talk at age about eleven, and his attempts to avoid going to school.

Christopher's behavior until that age involved more than apparent goose-stepping. Upon watching programs about Germany on television he would sometimes note a minor detail about a character's uniform that was incorrect, such as the badge on the uniform in the wrong place. Upon watching a documentary about the Holocaust, and seeing a scene at a concentration camp, that his parents thought might have been Auschwitz, Christopher said that his airplane base was near that camp.

Friends of the family would sometimes ask Christopher for details of his prior life, such as his clothing, all of which he would answer. They sometimes asked for sketches of something from his previous life. Christopher would oblige and the visitors would take the sketches home with the parents' permission. The parents however, unfortunately, kept none of the sketches Christopher made spontaneously.

Christopher said that his name in his prior life was Robert. His father was Fritz, and a brother was named Peter. His previous mother was dark and wore glasses, but he did not remember her name. He had also told his present parents that he was 23 when he was killed and that his fiancée, blond and thin, was 19.

Christopher was extremely blond, his hair was described as straw colored, and eyebrows and eyelashes were blond. His eyes were blue. Though his mother, Veronica, had blue eyes, all members of the immediate family, other than Christopher, had brown hair.

He differed from other members of his present family in many behavioral respects. Though the others liked to drink tea, he preferred thick soup and sausages. He expressed a wish to live in Germany. Once, he insisted on being given the part of a German in a play being given at school.

Where did it all come from? There seems little in the immediate family to arouse curiosity about a connection there. Stevenson invested much time and effort trying to track down some possibilities, all being dead-enders. Veronica's father fought with the British army in in Africa in World War II and had a hatred of Germans. By the mid-seventies there was little talk in the Ellis household about the war. Jack, Christopher's father, was confident Christopher was in bed and would not have witnessed any of the rare films shown on TV at night about the war. In later years he did witness a few of such films, long after he had begun talking and acting like a German.

The parents were members of the Church of England. They seemed uninterested in the subject of reincarnation and initially thought Christopher's behavior and statements amounted to a childhood phantasy. However they made no attempt to suppress any of it. Stevenson suggests that the father may have unconsciously and indirectly encouraged it by asking questions about it. After some years however they became convinced that he was referring to a real life he believed he had lived.

The family possessed no books about WW II. Jack told Stevenson, that at times he, Jack, would look at books from the library to check the accuracy of things his son had said, and that he was always found to be correct.

Jack's brother had married a German woman whose father had been a German Air Force pilot killed in the war. Contacted by Stevenson she refused to discuss her first husband. Stevenson learned that she knew of Christopher and his statements but "did not believe in it." Stevenson also learned that a German Dornier aircraft had crashed very near Middlesbrough, where Christopher was born, but nothing discovered about it jibbed with Christopher's account.

Stevenson wrote that this case demonstrates the need to investigate when the subject is still young, and completing the investigation as soon as possible. He states also that he has given as evidence in his report only what the parents said of Christopher's statements as a youngster. In 1993, the time of his investigation, 18-19 years after Christopher first spoke about a previous life, Stevenson felt there was too much risk that he may have added items either to please questioners, or from normally acquired Knowledge.

Stevenson concluded that there were enough details to conclude that the case could not be explained by "present knowledge of genetics or environmental influences." Therefore, he says, "I believe reincarnation is at least a plausible explanation for it."

It does not seem that any conclusion can validly be inferred from Stevenson's inability to find a likely candidate for the role of the previous personality. If there was indeed a viable candidate somewhere, it need not have been a relative nor close acquaintance. Nor should we be surprised if some details of his crash were found not be completely accurate. Nonetheless, Dr. Stevenson concludes only that reincarnation is "at least a plausible explanation" of this case.

Cases such as this might make us wonder if Stevenson has not set the bar too high for accepting a case as 'solved,' a term he uses only when the prior personality can be identified to a reasonable degree of certainty. In a number of cases it would seem almost impossible to find a prior personality.

They include among our thirteen, this one, the previous case of Daniel Lewis, and of Teuvo Koivisto, which we will examine next. Both of the prior personalities in those cases were obviously murdered in Nazi death camps. The same will be true, for a number of others. In all of these 'unsolved' cases reincarnation, as the term is used by Stevenson and others, appear as the only reasonable explanations despite the inability to identify prior personalities.

*   *   *

Teuvo Koivisto was the youngest of four children born to Jan Koivisto and his wife Lusa, in Helsinki, Finland. Teuvo's parents were Finnish as were his grandparents. One of his great-great grandmothers had been Polish and her mother was Jewish. During World War II Finland was allied with Germany in its invasion of Russia hoping to recover territory lost to Russia following the Soviet invasion of Finland in 1939. There were relatively few Jews in Finland and that country was highly uncooperative with German requests that it surrender those German Jews who found shelter there from Nazi persecution.

During the time of Lusa's pregnancy with Teuvo, she had a dream, possibly while only half asleep. In the dream she was standing in the line of prisoners which was moving forward, and someone said to her "Take shelter under the straw." She escaped and was confronted with a man who had a copy of the Kabballah, the text of a mystical school of thought originated in Judaism. Some men there were shooting. One said, "The baby you are expecting will be a Jew and I will save your life." With that the dream ended.

Teuvo was born on August 20, 1971. Otherwise healthy, he showed a fear of the dark early on, and his parents left a light burning where he slept at night. He began to speak at about 18 months, in full sentences when he was about 2, but not very fluently until about 3. During the 4th year he greatly surprised his mother by describing what it was like to be in a "concentration camp," though that word was not used, and there put to death with gas.

Rita Castrén, a professional caregiver, in what capacity Stevenson does not say, was informed of this case in early 1976 and on February 2nd of that year interviewed Lusa. She continued to act as interpreter with Stevenson on this case as she had on others, and sent a copy of her notes to him. It was December 1, 1978 before Stevenson was able to travel to Helsinki to interview Lusa. Jan took no part as Teuvo had never spoken to him about his ideas of a past life.

Teuvo had told Lusa that he had been alive before. Then he referred to the big furnace, this time apparently using the word. He said that the people were piled high, "higgledy-piggledy in layers in the furnace, some lying on top of others. He said that he had been taken to the "bathroom," and that personal items were taken from people in the bathroom, including eyeglasses and gold teeth. Then the people were undressed and put into the furnace. Gas came pouring out of some place in the walls, and he could not breathe. Teuvo said that he knew he was going to be put in the furnace. He said that he came to his mother after seeing the others put in there. He also described an oven with children in it. He then said to Lusa, "then I came to you. I was given here. Are you happy Mummy that I came to you?" Sometime later, sounding depressed, he said, "I was caught by the barbed wire Come and get me off."

At the meeting in 1976, Lusa told Rita Castrén that Teuvo often repeated his first statements in the morning upon awakening. He continued to speak about the memories for about a year and a half. About the age of three, when beginning to speak fluently, Lusa was surprised by the vocabulary he could use to describe his experiences, though it was still not adequate and had to use his hands to describe the shape of the furnace. She described her son as extremely frightened and terrified, causing her to tell him fairy tales to distract him. Teuvo did not lose his fear of the dark until about age 7.

Up to the age of two Teuvo did not want to wear clothes, even when going outside in cold weather. As a young child, he hid himself by knocking down walls separating rooms, something he could do as the family residence had walls thin enough for a child to break.

More serious, upon first describing his memories, it became difficult for him to breathe for 10-15 minutes. This occurred sometimes twice a week, sometimes not for three months. Medical examination showed he did not have asthma. This symptom was still present in 1978 when Stevenson interviewed Lusa. Otherwise, his health was excellent.

Lusa was firm in her opinion that Teuvo could not have acquired such information from any normal sources. He was rarely permitted to watch Television and never anything involving violence. Whether this was because of his apparent terror or because of a general belief about adverse effects of such programs on young children, we are not told. His parents and older brothers never discussed such things as concentration camps or gas chambers in the presence of Teuvo. Again we do not know whether such restraint was imposed because of the child's terrors and apparent memories, or simply was something that never came up.

During the time Teuvo talked about a previous life, the family was living in their own home. They had no social relationships with their neighbors, and Teuvo was shy and never talked with them. No grandparents were living with them.

Stevenson was an indefatigable researcher, but in this case he outdid himself. Not content with the obvious similarities between Teuvo's statements and what was common knowledge to most watchers of movies or TV (unrestrained and of an age of understanding), Stevenson armed himself with much of the scholarly research detailing the horror of the camps. Professional that he was, he allowed himself no show of emotion or judgment about the Nazi bestiality. But the thoroughness of the research, even by his own high standards, tells us all we need to know of it. Every one of his findings, described below is supported by scholarly reference.

Certain things were well known from the media, including barbed wire, seizure of personal property, forced undressing of the victims, the pretense by the Germans that the gas chambers were bathrooms even with signs to that effect, death from poison gas, and burning of the victims' bodies in furnaces, sometimes in open or pit fires. The Germans removed the gold teeth of the prisoners after they had killed them by gassing, but before the bodies were thrown into the fire.

Stevenson cites authority however for the fact that sometimes children at Treblinka and children and sometimes women were thrown into fire pits while still alive. Teuvo's mention of the prisoners' difficulty in breathing while being gassed is confirmed by the eyewitness account of a Jewish Hungarian physician, a Dr. Miklos Nyiszli arrested and sent to Auschwitz in April 1944. He was there selected by Dr. Josef Mengele to assist in medical experiments. There were a number of other prisoners with education and skills who were selected to help the SS, the group who controlled the camps. Most of them were later killed by the Nazis to suppress evidence of the grisly crimes. Dr. Nyiszli survived and wrote and published a full account of procedures used to kill Jews in the camps.

At Auschwitz they used a volatile preparation of Zyklon-B, which killed within 5 to 15 minutes. The prisoners were forced into a room and told to undress and leave all their clothing. The Germans thus implied they would need them later; in fact they were later used by the Germans. The rooms had signs indicating it was as bathroom. They were then crowded into a second room with upright pipes rising from the floor with holes in their sides. The doors were then closed and gas poured down the pipes to the room below and the gas then escaped through the holes in the upright pipes and poisoned the people in the death chamber.

Stevenson here quotes Nyiszli: "The gas first inundated the lower layers of air and rose but slowly to the ceiling. This forced the victims to trample one another in a frantic attempt to escape the gas." After the prisoners were all dead, their bodies were hauled out of the death chamber, without any of the gold teeth, then pushed into the crematoria whose furnaces dispelled through chimneys smoke and odors of burning flesh. The bodies of prisoners, according to Nyiszli, were piled up on top of each other. Other researchers have described the piling up of bodies in other death camps such as Treblinka and Dachau.

Barbed wire encircled all the camps. Three different authors were cited by Stevenson attesting to the details. The high outer wire conducted a lethal current of high voltage that immediately killed anyone coming in contact. No one could be taken off the barbed wire alive, such as when one came in contact with wire that was not electrified. At huge camps, including Auschwitz, unelectrified barbed wire separated different sections of the camp.

According to one of the authorities cited by Stevenson, like those too old to be of any use, the children under 14 were generally considered useless also and upon arriving at the camps went straight to the gas chamber or fire pits.

Stevenson makes a number of observations, comments, and observations. He states that the 'hiding behavior' possibly corresponds with activities in the Warsaw ghetto before the revolt of 1943, including preparations for it. He quotes one of his sources to the effect that the breaking of walls was very important in the plans of the Jews in the frantic movement of persons and supplies between rooms, apartments, cellars and attics, linking finally an entire residential block without going outside.

Stevenson is not certain that Teuvo was killed either at Auschwitz or the adjoining Birkenau camp. He quotes two of his sources to the effect that the Germans used similar methods of killing in other camps including Treblinka and Sobibor. At these latter camps, carbon monoxide of exhaust fumes from gasoline engines was the agent of death.

As he has in many other cases, Stevenson was interested in following up on the subject's later life, and talked with Teuvo in September 1999. He had graduated from a business school but chose music as a vocation and at the time of the interview was working as a professional musician and teaching music. He said that his difficulty in breathing ceased when he began school at age 5. He married in 1997, and by 1999 the couple had a 2 year old son. He said he had no 'imaged' memories of the previous life, but his hiding behavior lasted until age 13 or 14. He remembered from his earliest years he always wanted to feel safe and complained that his present residence had no hiding place.

He also claimed he felt anxiety when he saw Nazi uniforms or swastika flags, both of which caused him to stand still with fear, though when seeing the flags of England or France he felt no fear. He was interested in religions, but had no particular attraction to Judaism. Not surprisingly, he did believe in reincarnation, but believed it was incompatible with Christianity.

Stevenson's belief in the high probability, at least, of reincarnation here is attested to by his lack of highlighting any weak spots in the case. He almost never expresses a high level of belief, though many times such level, in spite of himself, exudes from his comments. This case is one of them.

Chapter 15:

SOMETHING SURVIVES

In the Epilogue to my book, *Our Quantum World and Reincarnation*, I said a number of things that require some modification, or perhaps in some cases, excision.

For one thing I had asked where the memories might be between the death of the prior personality and the birth of the subject. Except for the few cases of descriptions by the subject of the events occurring almost immediately following death of the prior personality,' accounts of that in-between period, whether of the 'other realm' or of Earthly matters are very scarce. Those few accounts are quite similar to those of 'near death experiencers,' a subject I have not dealt with in my book or in this essay. It is an interesting subject in itself.

There was no reason offered for this paucity in any literature I could find. I did in this essay offer what seems a plausible reason, the probable inability of separated atoms, entangled or not, to complete the necessary processes for encoding or retrieval of memories. Retrieval would come when the pertinent atoms would again be part of a living human. But with no encoding in the interim, there would be little or nothing to retrieve from that period.

I noted also in the Epilogue of my book that it was probable that not everyone's store of memories finds a new home in a new body, and asked whether they stayed, entangled or otherwise, in the vastness of space in perpetuity. I mention the possibility in this essay, as I did in the text of the book, the possibility that they became parts of other humans, animals or vegetation, or of rocks or stones.

I believe that the addition in this essay of summaries of the writings of Krauss and Levi should clear up any lack of explicitness in the book. The pertinent atoms could be in any of those places just mentioned, but considering the intricacy of the processes for encoding, no such atoms in a stone or animal or plant would make such repositories as stones alive, contrary to the beliefs of some animists, including physicist David Bohm and others. Nor, if existing in plants or animals other than humans, would it make them human.

Perhaps, however, the most important change should be the addition of the material in this essay in Chapter 12 concerning, not with what happens in entanglement, but *why* it happens. In my book, I stated (p. 65) that "Not many things are any longer unthinkable, but one of the few is the proposition that entanglement exists merely to keep intact some arcane characteristics of atoms, with no other consequences." The arcane characteristics were such things as spin and polarization. But for what purpose did they, or entanglement itself exist?

Fortunately it appears that some scientists are working on answers. In Chapter 12 of this essay I have attempted to explain two propositions involving functions of great importance that they advance. One is that it is entanglement that plays a key role in keeping the universe together, and that "space without entanglement could not hold itself together," and as stated by another "quantum entanglement is the needle that stitches together the tapestry of cosmic spacetime."

Hardly anything could be of greater importance. Not all scientists agree with that proposition however. Whether it will ultimately receive widespread acceptance remains to be seen.

A second suggestion by other scientists deals with the "arrow of time," the question of why energy flows continually toward equilibrium, and why it does not tend at times toward greater inequality, highs and lows, as the laws of classical physics would permit. Translated into a simpler example, why does a cup of hot liquid cool toward room temperature instead of raising the temperature of the room to its own heat level? By the laws of classical physics one should be as likely as the other.

But those laws did not take into account the process of entanglement, the combining of atoms of both the coffee and the room, and the consequent change toward equilibrium. There is obviously still much research to be done in this area, but the fact that the idea is spreading among physicists is indeed encouraging, though still without widespread enthusiasm among them.

In the Epilogue of my book I drew a parallel between these 'foreign memories' mostly in the very young, and AM radio that was used in those ancient days before World War II. The listeners often had to deal with interruptions from other broadcasts. Most important, think of the number of messages, both then and now, beyond our ability to see or hear, unless we are on the particular frequency for receipt of the message. There are for starters the wireless telephones, the cell phones, smart phones, I phones, and radio broadcasts. How many at any given time are in the 'ether', or whatever? Billions? Trillions? How many TV broadcasts, emails, faxes etc. share the same 'ether,' or space? More billions?

Yet we go about our business unconcerned about any of it, except those we choose, or those intended for us. Does this not make it less difficult to think of millions, or greater, of conglomerations of atoms with the potentiality of memories to be retrieved under the proper circumstances? The more I think of the amount of electromagnetism in our immediate space, the more efficacious I find the parallel between the imperceptible aural and visual messages, and atoms needing only, apparently by chance, a functioning body with which to communicate. I am no longer 98% uncertain about the comparison, as I said I was in the book Epilogue. I hesitate to choose what a new number would be.

I claimed in that Epilogue that there was only one thing of which we could be certain, namely that "We have thus far discovered only a very tiny fraction of a percent of the nature of our universe, and the laws that make it work." I had underestimated the progress our scientists were making. I again would not be so bold, or so foolish, as to estimate with any guess as to how far along we are, but it may be at least a bit more than I had thought; the ground has almost literally moved beneath our feet.

I said also that I imagine and believe that there are forces 'out there' of whose existence we are totally unaware. That is still true. We have become aware of new possibilities for forces of which we were already aware, but their place in the larger scheme of things is being explored and the tentative ideas could be as intriguing as that of any new forces. They could, except that we still do not know what we do not know.

I also said that we had only a tenuous foothold in this strange and unfamiliar world and that we still do not know what may lie beyond it. We still do not, but we can sense something of it, and even what we can merely sense is undoubtedly more startling than anything we might imagine.

One thing I would not change is the title of this essay. I have not, and do not claim that my hypothetical explanation of the cause of the foreign memories of some children is proven, that entanglement is in fact the mechanism that endows them with something that has survived from an earlier life.

What I do feel has been proved beyond reasonable doubt is that something does indeed survive, whatever the mechanism may be, at least in an unknown number of cases. That has been proved, not by me, but by the investigations of Dr. Ian Stevenson and others who have followed his lead. The effect of this 'third layer of inheritance' seems, insofar as the research shows, usually not to have been of profound or lasting impact on the subjects studied. But the fact that something does indeed survive could yet be shown to be of profound and lasting significance. We are only at the beginning of a new dimension to explore.

Other generations have had their conquerors and tyrants, still others their explorers and statesmen. But our future will be carved by science and the scientists, something I believe that makes our time, atrocities, widespread poverty, and endless slaughter notwithstanding, truly the most intellectually stimulating and the most challenging in all of human history thus far.

BIBLIOGRAPHY FOR

SOMETHING
SURVIVES

Aczel, A. E*ntanglement.* New York: Penguin, 2001.

Bache, C.M. *Life Cycles.* New York: Paragon House, 1991.

Bernstein, J. *Quantum leaps.* Cambridge, MA: Belknap Press, 2009.

Bohm, D. and Hiley, B.J. *The Undivided Universe.* New York: Routledge, 1993.

Clodd. E. *Animism.* London: Archibald Constable & Co, 1905.

Cox, B. and Forshaw, J. *The Quantum Universe.* Boston: Da Capo Press, 2011.

Davies, P. *The Cosmic Blueprint.* New York: Simon & Schuster, 1988."

Gilder, L. *The Age of Entanglement.* New York: Vintage Books, 2009.

Gisin, N. *Quantum Chance.* Heidelberg: Springer, trans. S. Lyle, 2014.

Graham, H. *Animism.* New York: Columbia UP, 2006.

Hardo, T. *Children Who Have Lived Before.* London: Random House, 2005 (First published in Germany by Verlag Die Silberschnur, 2000.

Hardo, T. *The Karma Handbook.* Mumbai: Jaico Publishing House, 2007 (Published in arrangement with Silberschnur Publishing House)

Hardo, T. *30 Most Convincing Cases of Reincarnation.* Mumbai: Jaico Publishing House, 2009. (Published in arrangement with Verlag "Die Silberschnur")

Herbert, N. *Elemental Mind.* New York: Penguin, 1993.

Herbert, N. *Faster Than Light.* New York: New American Library, 1988.

Holt, J. *Why Does the World Exist?* New York: Liveright Publishing, 2012.

Jablonka, E. and Lamb, M.J. *Epigenetic Inheritance and Evolution.* New York: Oxford UP, 1995.

Haraldsson, E. "Children Claiming Past-Life Memories," in Journal of Scientrjic E,rploratron, Vol.5, 1991, 235-243.

Haraldsson, E. *The Departed among the Living.* Guilddford, United Kingdom: White Crow Books, 2012.

Holcombe, L. *The Presidents and UFOs.* New York: St Martin's Press, 2015

Kaku, M. Hyperspace. New York: Anchor Books, 1995.

Krauss, L. M. *Atom.* Boston: Little Brown, 2001.

Levi, P. *The periodic Table.* New York: Schocken Books, R. Rosenthal, trans., 1984.

Magee, B. *The Tristan Chord.* New York: Metropolitan Books, 2000.

McDougall, W. *Body and Mind.* New York, 1911.

Mitchell, E. *The Way of the Explorer.* Franklin Lakes, NJ: New Page Books, 2008.

Nadeau, R. and Kafatos, M. *The Non-Local Universe,* New York: Oxford UP, 1999.

Pasricha, S. *Claims of Reincarnation,* New Delhi: Harman Publishing House, 1990.

Po, Huang. *The Zen Teaching of Huang Po.* Trans. John Blofeld. New York: Grove Press, 1959.

Rane, W. *Soul Hunters.* Berkeley, CA: University of California Press, 2007.

Rosenblum, B. and Kuttner, F. *Quantum Enigma.* New York: Oxford UP, 2011.

Schopenhauer, A. *The World as Will and Representation.* Vol. I and II. Trans. E.F.J.

Shroder, T. *Old Souls.*, New York: Simon & Schuster, 1999. Payne. New York: Dover Publications, 1966.

Seife, C. *Decoding the Universe.* New York: Penguin, 2006.

Skrbina, D. *Panpsychism in the West.* Cambridge, MA, 2007.

Spiro, M. E. *Burmese Super-Naturalism*, Expanded Ed. New Brunswick, NJ: Transaction Publishers, 1996.

Stevenson, I. *Children Who Remember Previous Lives.* Jefferson, NC: McFarland and Company, 2001.

Stevenson, I. *European Cases of the Reincarnation Type.* Jefferson, NC: McFarland and Company, 2003.

Stevenson, I. *Twenty Cases Suggestive of Reincarnation,* 2nd Ed, revised and enlarged. Charlottesville: University of Virginia Press, 1974

Talbot, M. *The Holographic Universe.* New York: Harper Collins, 1991.

Tucker, J. B. *Life Before Life.* New York: St. Martin's Press, 2005.

Tucker, J. B. *Return to Life.* New York: St. Martin's Press, 2013.

Turkheimer, E. and Gottesman, I. "Individual differences and the canalization of human behavior", in *Developmental Psychology,* Vol. 27(1), Jan 1991.

Tylor, E. B. *Anthropology.* New York: Appleton and Company, 1881.

Waddington, C.H. "Evolutionary Adaptation," in *Learning, Development and Culture.* Ed. H.C. Plotkin. Chichester, NY: John Wiley & Sons, 1982

Zohar, D. *The Quantum Self.* New York: William Morrow, 1990.

CPSIA information can be obtained
at www.ICGtesting.com
Printed in the USA
BVHW031942210921
617213BV00005B/122